International Conference on Engineering Materials (ICEM 2019)

IOP Conference Series: Materials Science and Engineering
Volume 838

Chengdu, China
25 - 27 October 2019

ISBN: 978-1-7138-2680-4
ISSN: 1757-8981

Printed from e-media with permission by:

Curran Associates, Inc.
57 Morehouse Lane
Red Hook, NY 12571

CURRAN ASSOCIATES INC.
proceedings
.com

Some format issues inherent in the e-media version may also appear in this print version.

Institute of Physics
Dirac House, Temple Back
Bristol BS1 6BE UK

Phone: 44 1 17 929 7481
Fax: 44 1 17 920 0979

techtracking@iop.org

Additional copies of this publication are available from:

Curran Associates, Inc.
57 Morehouse Lane
Red Hook, NY 12571 USA
Phone: 845-758-0400
Fax: 845-758-2633
Email: curran@proceedings.com
Web: www.proceedings.com

TABLE OF CONTENTS

MATERIAL PREPARATION AND CHARACTERIZATION

STRUCTURAL DESIGN AND ANALYSIS OF ENGINEERING MATERIALS

IOP Conf. Series: Materials Science and Engineering **838** (2020) 011002 doi:10.1088/1757-899X/838/1/011002

Conference Committees

International Advisory Committee

Prof. Mohamed Henini, University of Nottingham, UK

Prof. Galina Ilieva, University of Plovdiv, Bulgaria

Conference Co-chair

Prof. Xiaohong Zhu, Sichuan University, China

Conference Program Committees

Prof. Ying-Hao Eddie Chu, National Chiao Tung University, Taiwan

Prof. Guneik Jang, Chungbuk National University, Korea

Prof. Tjokorda Gde Tirta Nindhia, Udayana University, Indonesia

Local Organising Chair

Prof. Jiliang Zhu, Sichuan University, China

International Technical Committees

Prof. Anders Gåård, Karlstad University, Sweden

Prof. Aidy Ali, National Defence University of Malaysia, Malaysia

Prof. Geetesh Goga, K.C. College of Engg. & I.T., India

Prof. Dr. Osman ADIGUZEL, Firat University, Turkey

Assoc. Prof. Mohd Rozi Ahmad, Universiti Teknologi MARA, Malaysia

Assoc. Prof. S. Balasivanandha Prabu, Anna University, India

Assoc. Prof. Abbas Teimouri, Payame Noor University (PNU), Iran

Asst. Prof. Dr. Rafia Usman Khan, NED University, Pakistan

Dr. T. THEIVASANTHI, Kalasalingam University, India

Dr. Syed Bahari Ramadzan Syed Adnan, University of Malaya, Malaysia

Dr. Nirali Gondaliya, SVM institute of Technology, India

Dr. Muhammad Irfan-ul-Hassan, University of Engineering and Technology, Pakistan

Dr. Abid Ali Shah, International Islamic University, Pakistan

Dr. Arjon Turnip, Indonesian Institute of Sciences, Indonesia

Dr. Azmah Hanim Mohamed Ariff, Universiti Putra Malaysia, Malaysia

IOP Conf. Series: Materials Science and Engineering **838** (2020) 011002 doi:10.1088/1757-899X/838/1/011002

Dr. Shanmuga Priya Natesan, Siddaganga Institute of Technology, India

Dr. Ping Xiang, City University of Hong Kong, Hong Kong

Dr. I. D. G. Ary Subagia, Udayana University, Indonesia

Dr. Hendra Suherman, Universitas Bung Hatta, Indonesia

Dr. M. Shamim Miah, University of Asia pacific, Bangladesh

Dr. Hugo Silva, University of Minho, Portugal

Dr. Mozaffar Salimi, Malek Ashtar University of Technology, Iran

IOP Conf. Series: Materials Science and Engineering **838** (2020) 011003 doi:10.1088/1757-899X/838/1/011003

First name	Surname	Institution	E-mail address
Hongbo	Qin	Guangdong Welding Institute (China-Ukraine E.O. Paton Institute of Welding), China	qinhb@guet.edu.cn
Junxue	Zhang	Southeast University, China	zhangjunxue2017@foxmail.com
Fuyong	Hu	Tianjin University of Technology, China	yongye673411728@163.com
Kai	You	Hohai University, China	wskhyk4@163.com
Bo	Zhao	China Special Equipment Inspection and Research institute, China	zhaobo19840626@163.com
Ruining	Wang	Xi'an Polytechnic University, China	wangruining123@163.com
Fang	Liu	Northeastern University, China	liufang@smm.neu.edu.cn
Zhenyu	Zhang	Shenyang Aerospace University, China	285226827@QQ.COM
Kai	You	Hohai University, China	wskhyk4@163.com
Zhengmao	Cao	China Merchants Chongqing Communications Research and Design Institute Co., Ltd., China	swjtuczm@163.com
Feng	Chen	Shanghai University, China	qinfengd@sina.com
Wuxiao	Wang	Xi'an University of Technology, China	wangwuxiao@163.com
Fei	Gao	Northeastern University, China	gaof@mail.neu.edu.cn

IOP Conf. Series: Materials Science and Engineering **838** (2020) 011004 doi:10.1088/1757-899X/838/1/011004

Peer review statement

All papers published in this volume of *IOP Conference Series: Materials Science and Engineering* have been peer reviewed through processes administered by the proceedings Editors. Reviews were conducted by expert referees to the professional and scientific standards expected of a proceedings journal published by IOP Publishing.

IOP Conf. Series: Materials Science and Engineering **838** (2020) 012001 doi:10.1088/1757-899X/838/1/012001

The influence of temperature on the mechanical properties of polycrystalline and single-crystalline Ni_3Al

T F Kuang[1,2], H B Qin[1,2,*], F M Liu[1], Q Li[1], X Yue[1], H T Gao[1] and Y Y Yi[1,*]

[1] Guangdong Provincial Key Laboratory of Advanced Welding Technology, Guangdong Welding Institute (China-Ukraine E.O. Paton Institute of Welding), Guangzhou 510650, China
[2] Key Laboratory of Guangxi Manufacturing System and Advanced Manufacturing Technology, Guilin University of Electronic Technology, Guilin 541004, China

[*] Corresponding author: yiyy@gwi.gd.cn; qinhb@guet.edu.cn

Abstract. Because of the excellent mechanical property at high temperature, single-crystalline Ni_3Al-based alloys, which belong to Ni-based superalloys, are widely adopted in the fields of aerospace, aviation and military. Usually, single-crystalline Ni-based alloys are bonded or welded together by polycrystalline Ni-based alloys. In this study, the Young' s modulus, shear modulus, bulk modulus, and Poisson ratio of polycrystalline Ni_3Al at the temperature ranging from 10 °C to 850 °C were calculated according to Voigt - Reuss approximation, and the relationships between the mechanical properties and the temperatures were clarified. Moreover, the influence of temperature on the universal anisotropy index and mechanical properties of single-crystalline Ni_3Al was evaluated. Results show that, except for Poisson ratio, the increase of temperature linearly decreases the mechanical properties of polycrystalline Ni_3Al, while it increases the universal anisotropy index of the single-crystalline Ni_3Al. Moreover, different with the bulk modulus, the Young's modulus and shear modulus of single-crystalline Ni_3Al are obvious anisotropic, and their anisotropic characteristics are quite opposite. The increasing temperature decreases both maximum and minimum values of Young's modulus, shear modulus, and bulk modulus of single-crystalline Ni_3Al.

1. Introduction

For the excellent mechanical and physical properties at high temperatures, such as extremely high strength, creep limit, fatigue limit, corrosion and oxidation resistances, nickel(Ni)-based superalloys are widely used in the fields of aerospace, aviation and military [1-3]. As surface imperfections, grain boundaries inevitably weaken the mechanical and other properties of metal at high temperatures. Accordingly, the latest generation of Ni-based superalloys are single crystals. Because of the poor forming performance of single-crystalline Ni-based superalloys and the complicated shape of parts or components, single-crystalline Ni-based alloys (i.e., parent alloy) are usually bonded or welded together by polycrystalline Ni-based alloys (i.e., intermediate layer) [4-6]. The state of the crystallographic orientation in the single-crystalline parent alloy are totally different with that in the polycrystalline intermediate layer. Although some mechanical properties of polycrystalline Ni-based alloys have been measured by experimental measurement, the influence of temperature on the mechanical properties is still not investigated systematically [7-9]. In addition, because of crystal anisotropy, there are countless crystal orientations in a single crystal, and the mechanical properties of single-crystalline Ni-based alloys in three-dimensional space are impossible to be measured by experimental method. In the previous study [10], first-principles calculations based on density function theory (DFT) were employed to calculate the mechanical properties of alloy in all crystal orientations, and the calculation results could be verified by experimental data. However, for the limitation of DFT, the influence of temperature on the mechanical properties of single-crystalline alloys can not be investigated via first-principles calculations. The authors' latest study shows that the differences of the mechanical properties induced by different states of crystal orientations between the single-crystalline parent alloy and polycrystalline intermediate layer induce stress and strain concentration in the bonding or welded joints [11]. Thus far, the in-depth and systematic comparative studies of single-crystalline and polycrystalline Ni-based superalloys are very scarce, especially for these at various temperatures.

Ni_3Al is a kind of intermetallic compound. Ni_3Al-based alloys, which have outstanding mechanical and physical performances, belong to the Ni-based superalloys. In the present study, the mechanical properties of single-crystalline and polycrystalline Ni_3Al were calculated, and the influence of temperature on the mechanical properties of polycrystalline and single-crystalline Ni_3Al was clarified.

2. Computational methods

Intermetallic compound Ni_3Al has a face-centered cubic lattice structure and *Pm-3m* space group, and its lattice constant is *a*= 3.572 Å, as shown in Figure 1(a) [12]. Considering the lattice symmetry of cubic structure, the stiffness matrix of single-crystalline Ni_3Al can be expressed as:

$$\begin{pmatrix} \sigma_1 \\ \sigma_2 \\ \sigma_3 \\ \tau_1 \\ \tau_2 \\ \tau_3 \end{pmatrix} = \begin{pmatrix} C_{11} & C_{12} & C_{13} & 0 & 0 & 0 \\ & C_{22} & C_{23} & 0 & 0 & 0 \\ & & C_{33} & 0 & 0 & 0 \\ & & & C_{44} & 0 & 0 \\ & Sym. & & & C_{55} & 0 \\ & & & & & C_{66} \end{pmatrix} \begin{pmatrix} \varepsilon_1 \\ \varepsilon_2 \\ \varepsilon_3 \\ \gamma_1 \\ \gamma_2 \\ \gamma_3 \end{pmatrix}$$

Where σ_i, τ_i, ε_i, γ_i are the normal stress, shearing stress, and the corresponding normal strain and shearing strain, respectively. The inverse matrix of stiffness matrix $[C_{ij}]$ is the flexibility matrix $[S_{ij}]$, and there are three independent variables (C_{11}, C_{44}, C_{12}, and S_{11}, S_{44}, S_{12}) in them, respectively, because of the lattice symmetry of cubic structure. The values of the independent variables ranging from 10 °C to 850 °C can be obtained from a previous study [13]. For a cubic lattice, according to the Voigt - Reuss approximation [14], the upper limits of bulk modulus (B_V) and shear modulus (G_V) can be calculated by Equation (1) and Equation (2), respectively:

$$B_V = \frac{1}{3}(C_{11} + 2C_{12}) \tag{1}$$

$$G_V = \frac{1}{5}(C_{11} - C_{12} + 3C_{44}) \tag{2}$$

The lower limits of bulk modulus (B_R) and shear modulus (G_R) are written as Equation (3) and Equation (4), respectively:

$$B_R = \frac{1}{3}(C_{11} + 2C_{12}) \tag{3}$$

$$G_R = 5(C_{11} - C_{12})C_{44} / [4C_{44} + 3(C_{11} - C_{12})] \tag{4}$$

The practical bulk modulus (B) and shear modulus (G) are estimated as the arithmetic mean, as given in Equation (5) and Equation (6), respectively:

IOP Conf. Series: Materials Science and Engineering **838** (2020) 012001 doi:10.1088/1757-899X/838/1/012001

$$B = \frac{1}{2}(B_R + B_V) \tag{5}$$

$$G = \frac{1}{2}(G_R + G_V) \tag{6}$$

Then, Young's modulus (E) and Poisson ratio (v) can be calculated by Equation (7) and Equation (8), respectively:

$$E = \frac{9BG}{3B + G} \tag{7}$$

$$v = \frac{3B - E}{6B} \tag{8}$$

The elastic anisotropy of single-crystalline Ni_3Al can be described by the universal anisotropy index A^U [15], which can be expressed as

$$A^U = 5\frac{G_V}{G_R} + \frac{B_V}{B_R} - 6 \tag{9}$$

To further understand the anisotropic mechanical properties of single-crystalline Ni_3Al, the Young's modulus E and bulk modulus B in all crystal orientations can be expressed as [16]

$$\frac{1}{E} = S_{11} - 2\left(S_{11} - S_{12} - \frac{S_{44}}{2}\right)\left(l_1^2 l_2^2 + l_2^2 l_3^2 + l_1^2 l_3^2\right) \tag{10}$$

$$\frac{1}{B} = (S_{11} + 2S_{12})\left(l_1^2 + l_2^2 + l_3^2\right) \tag{11}$$

Where l_1, l_2, and l_3 denote the direction cosines with respect to the a, b, and c axis (i.e., directions [100], [010], and [001]) of the lattice. The shear modulus G in all crystal orientations can be expressed as Equation (12):

$$\frac{1}{G} = (S_{44} + 4S_0 J), \text{ and } J = \sin^2\theta \cdot \cos^2\theta + 0.125 \cdot \sin^4\theta(1 - \cos 4\varphi), \tag{12}$$

Where θ and φ are Euler angles in the direction cosines, see Figure 1(b).

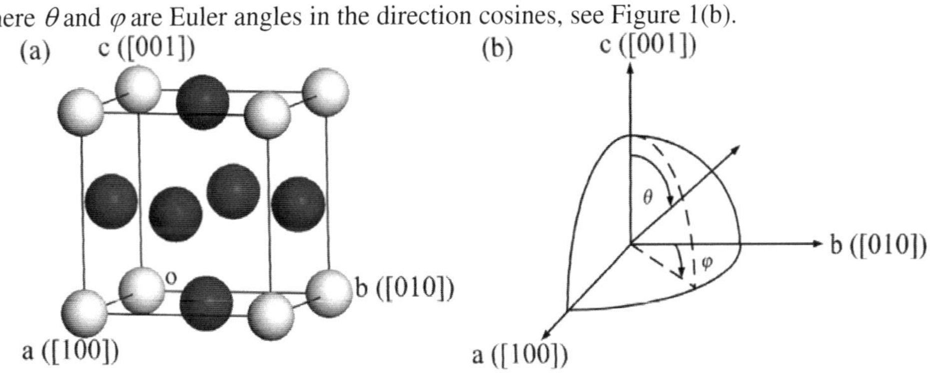

Figure 1. The structures of cubic Ni_3Al: (a) unit cell; and (b) coordinates system and Eulerian angles.

3. Results and discussion

Figure 2 shows the temperature dependences of the mechanical modulus and Poisson ratio of the polycrystal Ni_3Al at various temperature. Clearly, the Young's modulus, shear modulus, and bulk modulus of the polycrystalline Ni_3Al decrease linearly with the increase of temperature, see Figure 2(a). By linear fitting, the relations between these moduli and the temperature (T) can be expressed as:

$$E = 1.92832\times10^{11} - 5.08373\times10^{7}T\text{ , }R^2\text{=99.8} \tag{13}$$

$$G = 7.8808\times10^{10} - 2.16755\times10^{7}T\text{ , }R^2\text{=99.8} \tag{14}$$

$$B = 1.16147\times10^{11} - 2.4362\times10^{7}T\text{ , }R^2\text{=98.9} \tag{15}$$

Generally, Poisson ratio rises with the increasing temperature. However, different with the Young's modulus, shear modulus and bulk modulus, there is no obvious linear relation, see Figure 2(b). For example, the Poisson ratio will decrease obviously when the temperature is higher than 750 °C.

The arrangement characteristics of atom determine the mechanical anisotropy of material, which has a significant influence on the mechanical behaviour, e.g., crack growth, anisotropic deformation and elastic instability. Figure 3 exhibits the temperature dependence of universal anisotropy index in the single-crystalline Ni_3Al. Clearly, increasing temperature linearly increases the universal anisotropy index A^U of the single-crystalline Ni_3Al, which can be expressed as Equation (16) by linear fitting:

$$A^{U} = 1.61426 + 2.74628\times10^{-4}T\text{ , }R^2\text{=99.9.} \tag{16}$$

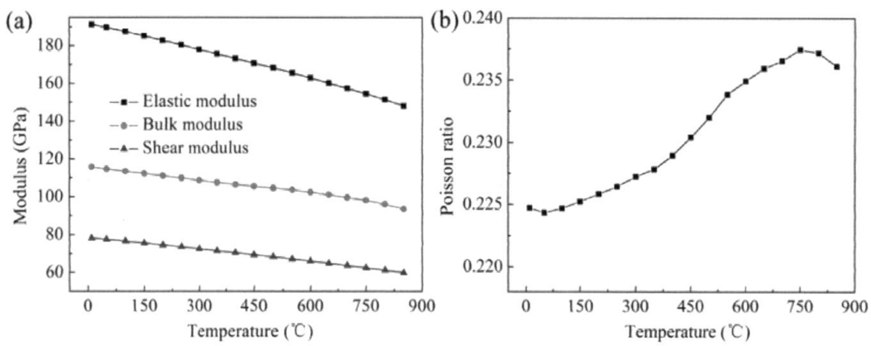

Figure 2. The temperature dependences of mechanical modulus (a) and Poisson ratio (b) of the polycrystalline Ni_3Al.

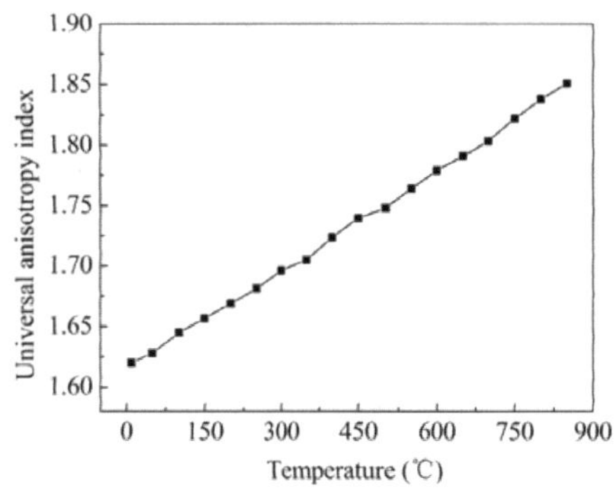

Figure 3. The temperature dependence of universal anisotropy index in the single-crystalline Ni_3Al.

Orientation dependences of Young's modulus in the single-crystalline Ni_3Al are illustrated in Figures 4(a1)–(a3), where in the maximum Young's moduli are in the directions <111> and the minimum Young's moduli are along the three axes (i.e., <100> directions) at various temperatures. Moreover, Figures 4(b1)–(b3) reveal that, contrary to the directional dependence of Young's modulus, the values of all shear modulus along directions <100> are maximal, while the minimum shear moduli are all located at directions <111>. Therefore, both the Young's modulus and shear modulus of single-crystalline Ni_3Al show obvious anisotropic characteristics, and they are quite opposite. Nevertheless, there is no anisotropic phenomenon for bulk modulus, i.e., the value of bulk modulus in all orientations is same at a given temperature, and the maximum bulk modulus equals to the minimum bulk modulus, see Figures 4(c1)–(c3). Figure 5 shows the influence of temperature on the maximum and minimum modulus of Young's modulus, shear modulus, and bulk modulus. Approximatively, both maximum and minimum values decrease linearly with the increase of temperature, and the maximum and minimum values at 10 °C, 400 °C and 850 °C are listed in Table 1.

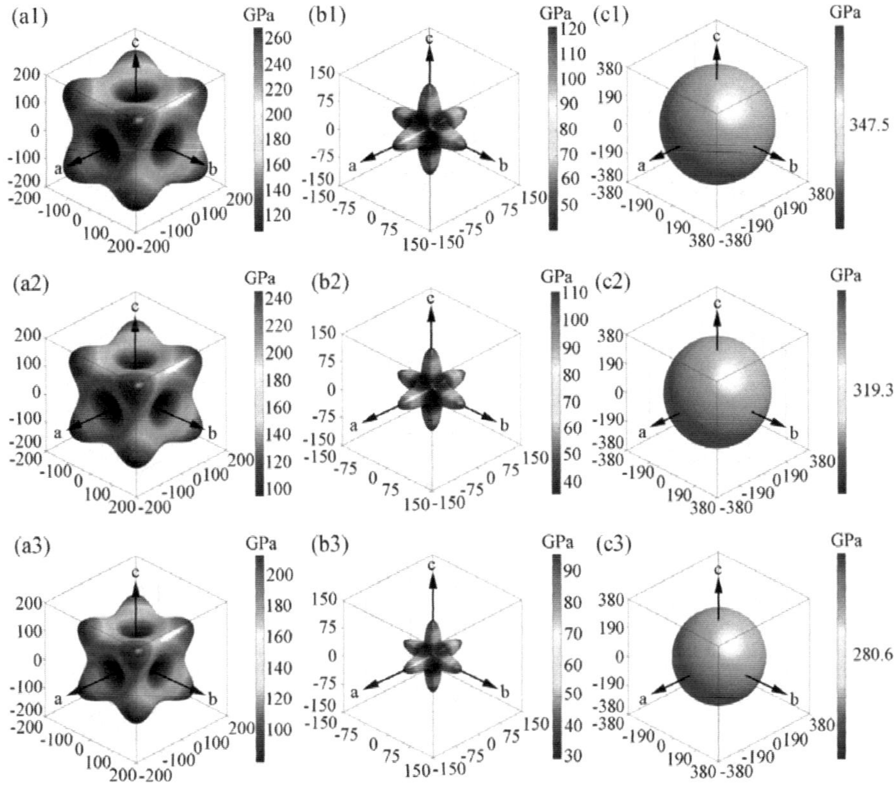

Figure 4. Crystal direction dependences of mechanical property in the single-crystalline Ni_3Al: (a1) Young's modulus, 10 °C; (a2) Young's modulus, 400 °C; (a3) Young's modulus, 850 °C; (b1) shear modulus, 10 °C; (b2) shear modulus, 400 °C; (b3) shear modulus, 850 °C; (c1) bulk modulus, 10 °C; (c2) bulk modulus, 400 °C; (c3) bulk modulus, 850 °C.

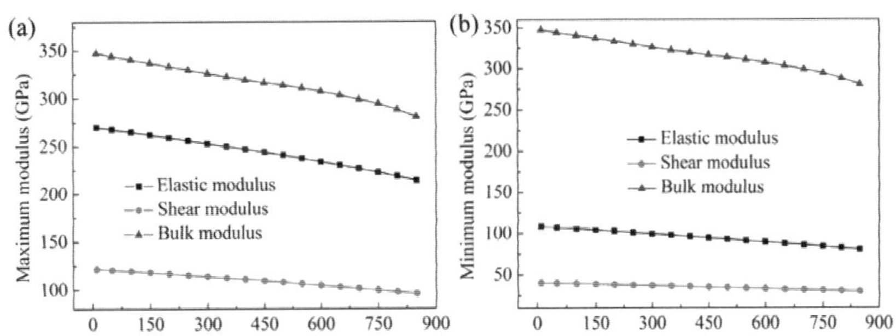

Figure 5. Temperature dependences of the maximum modulus (a) and minimum modulus (b) in the single-crystalline Ni_3Al.

Table 1. The maximum and minimum values of Young's modulus, shear modulus, and bulk modulus in the single-crystalline Ni_3Al at 10 °C, 400 °C and 850 °C.

Modulus	Maximum value (GPa)	Minimum value (GPa)	Temperature (°C)
Young's modulus	270.0	108.2	10
	246.9	96.1	400
	214.0	80.3	850
Shear modulus	121.5	40.3	10
	110.9	35.6	400
	95.7	29.6	850
Bulk modulus	347.5	347.5	10
	319.3	319.3	400
	280.6	280.6	850

4. Summary
To investigate the stress and strain concentration induced by different states of crystal orientations between the single-crystalline and polycrystalline Ni_3Al under different temperatures, in this study, the mechanical properties of both single-crystalline and polycrystalline Ni_3Al were evaluated and the influence of temperature on them were revealed. The Young's modulus, shear modulus and bulk modulus of polycrystalline Ni_3Al decrease linearly with the increase of temperature, and the relations between these moduli and temperatures have been established. Increasing temperature increases the universal anisotropy index of the single-crystalline Ni_3Al. The bulk modulus of single-crystalline Ni_3Al is isotropic. However, the Young's modulus and shear modulus are obvious anisotropic, and their anisotropic characteristics are opposite. The increasing temperature decreases both maximum and minimum values of Young's modulus, shear modulus, and bulk modulus of single-crystalline Ni_3Al.

5. References
[1] Caron P and Khan T 1999 *Aerosp. Sci. and Technol.* **3** 513
[2] Zhang P, Hu C, Ding C G, Zhu Q and Qin H Y 2015 *Mater. Design* **65** 575
[3] Smith T M, Rao Y, Wang Y, Ghazisaeidi M and Mills M J 2017 *Acta Mater.* **141** 261
[4] Paulonis D F, Duvall D S and Owczarski W A 1972 U.S. Patent 3 678 570
[5] Cook G O and Sorensen C D 2011 *J. Mater. Sci.* **46** 5305
[6] Khakian M, Nategh S and Mirdamadi S 2015 *J. Alloys Compd.* **653** 386
[7] Shah D M 1983 *Scripta metal* **17** 997
[8] Wang W, Jiang C B and Lu K 2003 *Acta Mater.* **51** 6169
[9] Wen Z, Zhang D, Li S, Yue Z and Gao J 2017 *J. Alloys Compd.* **692** 301
[10] Luan X H, Qin H B, Liu F M, Dai Z B, Yi Y Y and Li Q 2018 *Crystals* **8** 307
[11] Qin H B, Kuang T F, Li Q, Yue X, Gao H, Liu F M and Yi Y Y 2019 *Materials* **12** 2765
[12] Rao P M, Suryanarayana S V, Murthy K S and Naidu S N 1989 *J. Phys.: Condens. Mater.* **1** 5357
[13] Dickson R W, Wachtman J J B and Copley S M 1969 *J. Appl. Phys.* **40** 2276
[14] Hill R 1952 *Proceedings of the Physical Society. Section A* **65** 349
[15] Ranganathan S I and Ostoja-Starzewski M 2008 *Phys. Rev. Lett.* **101** 055504
[16] Zhang J M, Zhang Y, Xu K W and Ji V 2007 *J. Phys. Chem. Solids* **68** 503

Acknowledgments
This research was funded by Implementing Innovation-Driven Development Capacity Building Special Funds project of Guangdong Academy of Sciences (2018GDASCX-1005&2018GDASCX-0113), and Key Program for International Cooperation of Science and Technology (2015DFR50310).

Self-healing performance of intelligently building concrete module

J X Zhang

School of architecture, Southeast University, Nanjing 210096, PR China
Email: zhangjunxue2017@foxmail.com

Abstract. Intelligent building concrete module is an important part of parametric construction and its self-repairing performance is the premise condition for ensuring the quality of the building. In this paper, some experiments have been done using fly ask, slag, meta-kaolin and silica fume mixed into cement in order to test several performances, including compressive performance, impermeability and self-healing performance. The results demonstrate: the worst healing effect was fly ash, followed by slag, meta-kaolin and silica fume. In the cement-based materials, the self-repair of the cracks mainly concentrated in the early production of cracks, so paid attention to the protection of the early maintenance is the key for self-repair effect.

Keywords: Smart concrete material; Self-healing performance; Impermeability; Modular construction

1. Introduction

Cementitious materials, especially for the concrete, as the most used in building widely, have its unique features including heterogeneous components and different micro-environments that lead to cracks. The crack problem based-cement material is a quite important for the building, so it has a long history about the research in order to solve the difficulty.

There are two directions about crack treatment, one is a passive way and the other one is active style. For the passive way of repair work, it is a procedure of detecting, monitoring, and repairing through external conditions. The cracks are been detected, at the same time the repair work will be performing and the healing components penetrate into the cracks from the outside in order to fulfill the defects.

The earliest study of self-healing capacities about cementitious materials was in France and at that time scientists began to study the difficult problem using scientific method [1].

Cement-based building materials have several crack types and there will have different methods to deal with the problem in terms of the crack type. Cement-based materials have self-healing capability to block the crack generally, however, the capability is limited, cement-based composite materials become the object of study in this paper because of the self-healing capability of composite materials have a stronger than single cement material.

Using an electrodeposition technique to improve the crack problem of reinforced concrete had identified by Japanese researchers in the late 1980s [2]. Compared to traditional repair methods, the electrodeposition technique method has an obvious advantage by its effectiveness and low price [3]. Homma et al. [4] demonstrated the self-healing mechanisms of concrete materials, producing C-S-H swell to block the cracks in order to heal the concrete materials.

IOP Conf. Series: Materials Science and Engineering **838** (2020) 012002 doi:10.1088/1757-899X/838/1/012002

Yang [5] pondered the cracks of concrete from another direction that included influence of some essential environmental elements.

There were many reasons which can lead to the cracks of concrete, from saline water, acid rain and carbon dioxide of exterior elements including to interior causes, involved Reaction mechanism [6].

In conclusion, there are some methods to study the repair problem about concrete cracks, having the passive way and the active way through microencapsulation of intelligent materials, hydration of cement, expansive precipitations, bacterial method, and so on. In this paper, the self-repairing properties of composites are the focus of the study mixing with fly ash, slag, meta-kaolin and silica fume.

2. Experimental

The chemical analysis of cement was listed in Table 1. For Fly ash and Silica fume, their most compositions were SiO_2, 54.38% and 94.74%, respectively.

Table 1. Chemical composition of cement (%)

Composition	SiO2	Al2O3	Fe2O3	CaO	MgO	SO3	Na2O	K2O	L.O.I
Content	20.1	5.4	3.7	57.6	3.4	1.2	0.4	0.6	4.2

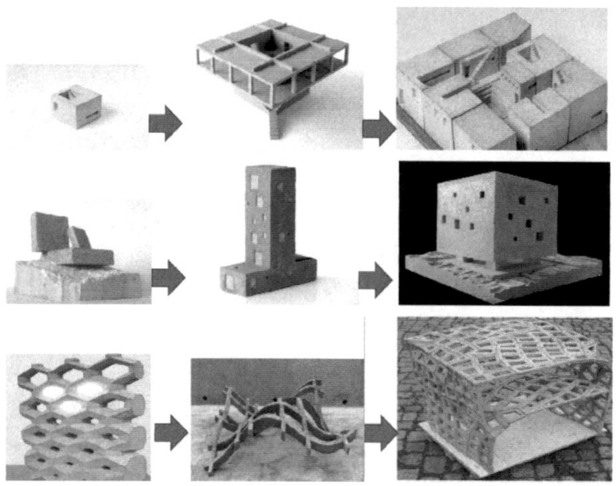

Figure 1. Parametric design type of concrete materials.

All specimens were produced for tests including compression degree, conservation time, admixture, chloride ion penetration to evaluate the capability of self-healing, as be shown in figure 1.

3. Results and discussion

3.1. Strength Recovery effect of different composite admixtures

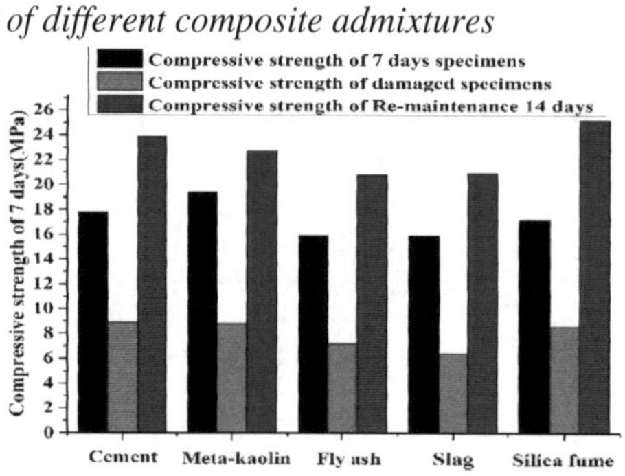

Figure 2. Effect of different composites on strength recovery.

Cement specimens of already damaged had shown a degree of self-healing effect to some extent under 14 days of standard conservations in fig.2. The phenomenon illustrates cemented materials specimens had self-healing capability duo to Continuous hydration process of unhydrated cement composition result in lots of gel product repair cracks. The hydration products mainly consists of $CaCO_3$ to block the cracks, which generated by $Ca(OH)_2$ and CO_2. The worst self-healing effect is the fly ash cement compared to Meta-kaolin specimen, slag specimen and silica fume specimen. The best effect of specimens be measured by compressive equipment is Meta-kaolin composite specimen, about 19.4MP after 7days conservation.

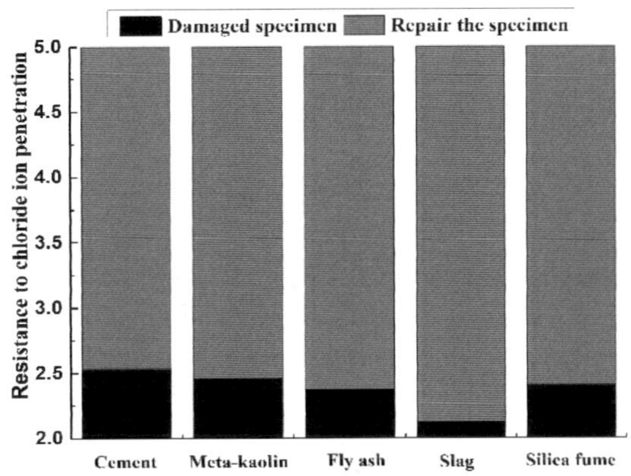

Figure 3. The rhythm of resistance of chloride ion penetration in concrete specimen.

Different composites impact the Resistance of chloride ion penetration in concrete specimen. From Fig.3, the data shows cement specimens have a weaker capability to resist the penetration of chloride ion than others' composites, whereas the capability to resist the chloride ion will increase obviously after 14 days conservation. In terms of micro-crack theory, the crack will gradually expand under the action of tensile stress, so the resistance of chloride ion penetration can improve along with the repair of cracks.

3.2. The influence of maintenance time for self-healing capability

As seen in Fig.2 and Fig.3, the self-healing effect of cement-based composite materials has different show. The strength of specimens with 7days maintenance time has obviously reduced than before.

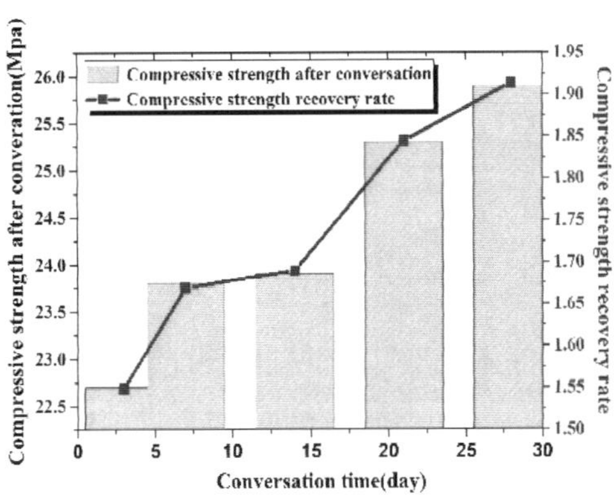

Figure 4. Effect of curing time on compressive strength

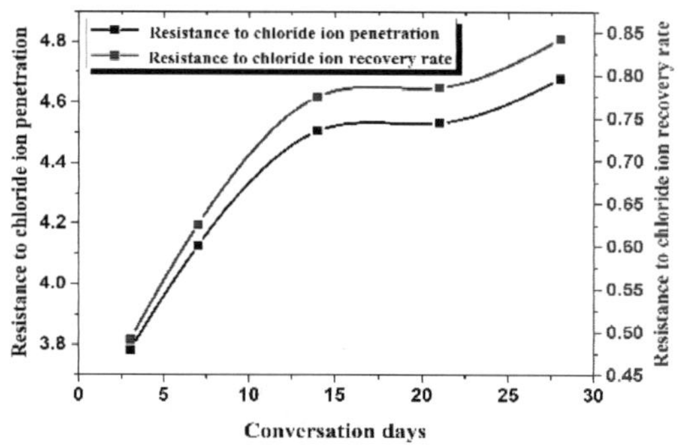

Figure 5. Compressive strength of the specimen.

With the increase of the curing time, the strength of the test specimens increased and became gentle, especially the early strength recovery effect was significant. Fig.4 presents the specimens of 3days conservation can be restored to 22.7MPa, amount to the same age non-pressure compressive strength of 95.38% The strength of the recovery is gradually stabilized, 28 days of strength back to 25.9MPa, having the same age of non-pressure compressive strength of 82.75%.

The results of the study are fully explained the cement-based material can be self-healing after be subjected to certain damage, and the self-healing phenomenon is sustained with the increase of the curing age, but the self-healing effect is obvious in the early stage from Fig.5.

Self-healing effect of cement-based materials is closely related to the hydration process and the hydration products are produced and the cracks inside the specimen are rapidly healed during the early stage of curing. With the deepening of the hydration process, the structure is gradually dense. Because water becomes more and more difficult to enter the cracks of specimen, it is not conducive to play from the healing effect further. Duo to the cover of the hydration product, the un-hydrated cement particles are prevented result in the hydration of cement hydration slowed down and cement-based material self-healing progress gradually weakened

4. Conclusion

The work aims to verify the self-healing of cement- based composite materials. There are some self-healing characteristics for all cement-based materials, but the repair scales are different for them.

A. A small amount of admixture has little effect on the self-repair of cement-based materials.

B. Adding fly ash and other admixtures, the self-healing effect of cement-based materials will be reduced, the worst healing effect is fly ash, followed by slag, meta-kaolin and silica fume.

C. The self-repair of cement-based materials mainly concentrated in the early stage.

5. References

[1] Lauer K R and Slate F O 1956 Autogenous healing of cement paste, *ACI Journal*, Proceedings. **52** 1083-97.

[2] Sasaki H and Yokoda M Repair method of marine reinforced concrete by electrodeposition technique, *Proceedings of the Annual Conference of Japanese Concrete Institute* 849–854.

[3] Yokoda M and Fukute T 1992 Rehabilitation and protection of marine concrete structure using electrodeposition method, *Proceedings of the International RILEM/CSIRO/ACRA Conference on Rehabilitation of Concrete Structures* 213–222.

[4] Homma D, Mihashi H and et al 2009 Self-healing capability of fiber reinforced cementitious composites, *Journal of Advanced Concrete technology*. **7** 217-228.

[5] Yang E H 2008 Designing added functions in engineered cementitious composites, Ph.D. Thesis, University of Michigan, Department of Civil Engineering, Michigan, USA.

[6] Snoeck D and De Belie N 2012 Mechanical and self-healing properties of cementitious composites reinforced with flax and cottonised flax, and compared with polyvinyl alcohol fibres, *Biosystems Engineering*. **111**(4) 325-335.

Acknowledgments

The research presented in this paper was supported by Postgraduate Research & Practice Innovation Program of Jiangsu Province (KYCX18_0104).

Effect of tungsten on creep properties of martensitic 9Cr steel

F Y Hu[1,2], S J Zhang[1,2], J J Shen[1,2,3]

[1]Tianjin Key Laboratory for Advanced Mechatronic System Design and Intelligent Control, School of Mechanical Engineering, Tianjin University of Technology, China
[2]National Demonstration Center for Experimental Mechanical and Electrical Engineering Education (Tianjin University of Technology), China

Corresponding author's e-mail address: sjj1982428@sina.com

Abstract. The effects of tungsten on the creep behavior and microstructure of tempered martensite 9Cr steel with different W concentrations ranging from 0–2.5 wt% were studied. The matrix consisted of tempered lath martensite structure and $M_{23}C_6$ located along the grain boundary or inside the grain. As the W concentration increased, the number of $M_{23}C_6$ increased while the particle size decreased. Also, with the increasing W concentration, the minimum creep rate and elongation decreased while the time to rupture increased. The rise in W concentration led to the refinement of $M_{23}C_6$, which enhanced the creep strength but reduced the creep plasticity.

Keywords: $M_{23}C_6$, the minimum creep rate, tungsten, the creep strength

1. Introduction

The 9–12% Cr tempered martensite ferritic steels are important high-temperature materials for fossil-fired power plants [1-4]. The excellent creep properties of these steels can be attributed to the stability of the grain boundaries and the martensite lath. The martensite lath is wrapped in the prior austenite grains (PAG) and contain high density of dissociated dislocations and dispersion of secondary phase particles [3,5,6,7]. $M_{23}C_6$ carbide is a chromium-rich compound, and the primary elements are Fe, C, Cr, Mo, and W [7]. $M_{23}C_6$ carbides mainly exist in the raw materials and are located along the grain boundaries and subgrain boundaries [7]. $M_{23}C_6$ carbides exert high zener resistance to inhibit the coarsening of martensite laths and dislocations movement. Also, they hinder the reaction between the dislocations and subgrain boundaries; thereby, can effectively enhance the creep strength [8-11]. However, $M_{23}C_6$ carbides tend to become coarser in long-term service, which may lead to decrease in long-term creep strength [7]. Therefore, long-term creep strength can be improved by slowing the coarsening rate of $M_{23}C_6$ carbide. The studies [12-17] have shown that component diffusion can control the growth and coarsening of $M_{23}C_6$ carbide in the ferritic heat-resistant steel. The addition of B to 9Cr heat-resistant steel can improve the high temperature stability of $M_{23}C_6$ [12-14]. Addition of Co can inhibit the coarsening of $M_{23}C_6$ during tempering [15-16]. Abe [17, 18] suggested that the coarsening speed of tungsten-containing $M_{23}C_6$ carbide is slower due to the low diffusion coefficient of W and the creep behavior is also affected by the W concentration. In this work, the effect of W content on the creep properties of 9%Cr ferritic heat-resistant steel was investigated to optimize the W content in 9%Cr ferritic steel.

Table 1. Chemical composition (wt.%) of 9% Cr steel.

	Mn	C	Si	Cr	Mo	V	Ni	Fe	P	S
0W	0.4	0.10	0.28	9.02	0.5	0.2	0.2	89.3	<0.001	<0.005
1.5W	0.4	0.10	0.21	8.95	0.5	0.2	0.2	87.8	<0.001	<0.005
2.5W	0.4	0.11	0.20	9.02	0.5	0.2	0.2	89.3	<0.001	<0.005

2. Experimental

The chemical composition of the examined steels is presented in Table 1. Only concentration of W was varied from 0 to 2.5 wt% in the steels, while the other alloying elements were kept constant (9%Cr, 0.1%C, 0.2Si, and 0.4Mn). These three steels are referred to as 9Cr0W, 9Cr1.5W, and 9Cr2.5W steel, respectively, depending on the W concentration. The required raw materials were melted in a vacuum induction furnace and then forged into three cylindrical materials of 50 mm in diameter and 160 mm in height. A 5 mm cube was cut and normalized at 1050°C for 165 min and 760°C for 136 min. The sample for microstructure observation was mechanically polished with a sandpaper and polishing agent, and then, etched with 10 ml of hydrochloric acid and 1 g of picric acid solution in 100 ml of ethanol. Optical microscope (OM, OLYMPUS-BX51M), Scanning electron microscope (SEM, ZEISS Gemini500), and Electron backscatter diffraction (EBSD) were utilized for the microstructural observation. The crystal morphology, $M_{23}C_6$ basic morphology, volume fraction, and dimensional change were studied. Coil spring specimens and conventional rectangular specimens were prepared using machine tooling (see Figure 1, 2). The tensile creep tests (RJ-50, 923 K, 120MPa) were conducted on the samples with different compositions of W, namely 0%, 1.5%, and 2.5%. Considering the low strain resolution of the traditional uniaxial creep experiments, the short-term creep experiments were carried out at 30 MPa and 923 K using the spiral spring creep test method [19-22]. The spiral spring creep test piece is shown in figure 2. In the spiral spring creep test method, torsion is the dominant factor in deformation [23]. As the stress or strain of the coil spring is essentially the shear component, it can be converted to an equivalent parameter using the Von-Mises equation [19]. The short-term creep experiment time was 270 ks.

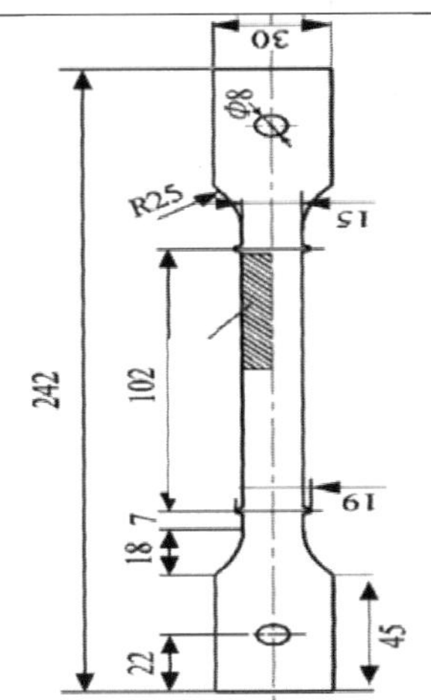

Figure 1. Rectangular specimens for uniaxial creep test.

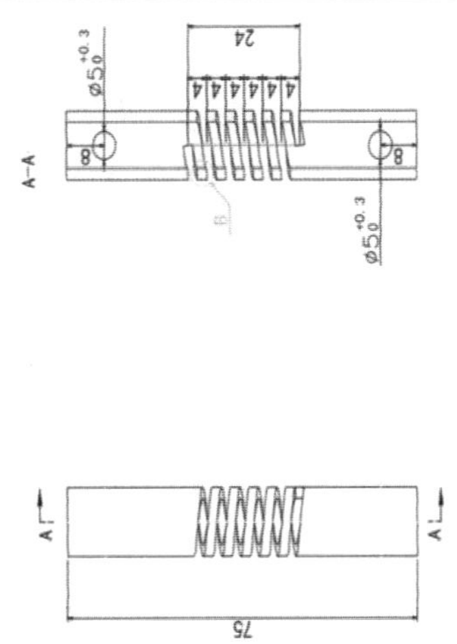

Figure 2. Helical spring specimens for creep testing under low stresses.

IOP Conf. Series: Materials Science and Engineering **838** (2020) 012003 doi:10.1088/1757-899X/838/1/012003

3. Results and discussion

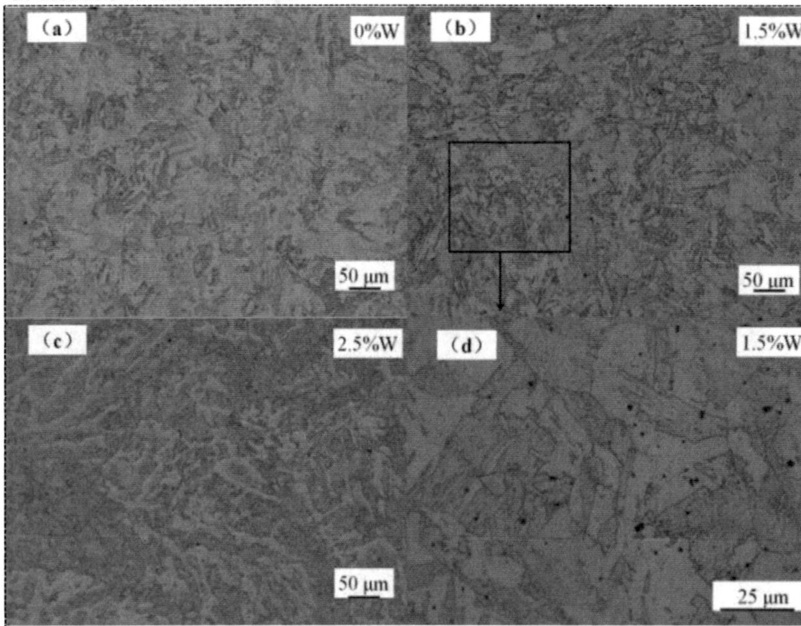

Fig. 3 Optical microscopy (OM) images of 9%Cr Heat-resistant Steel without aging. (a) 9Cr-0W, (b) 9Cr-1.5W, and (c) (d) 9Cr-2.5W, and (d) partial enlarged drawing, 9Cr-1.5W.

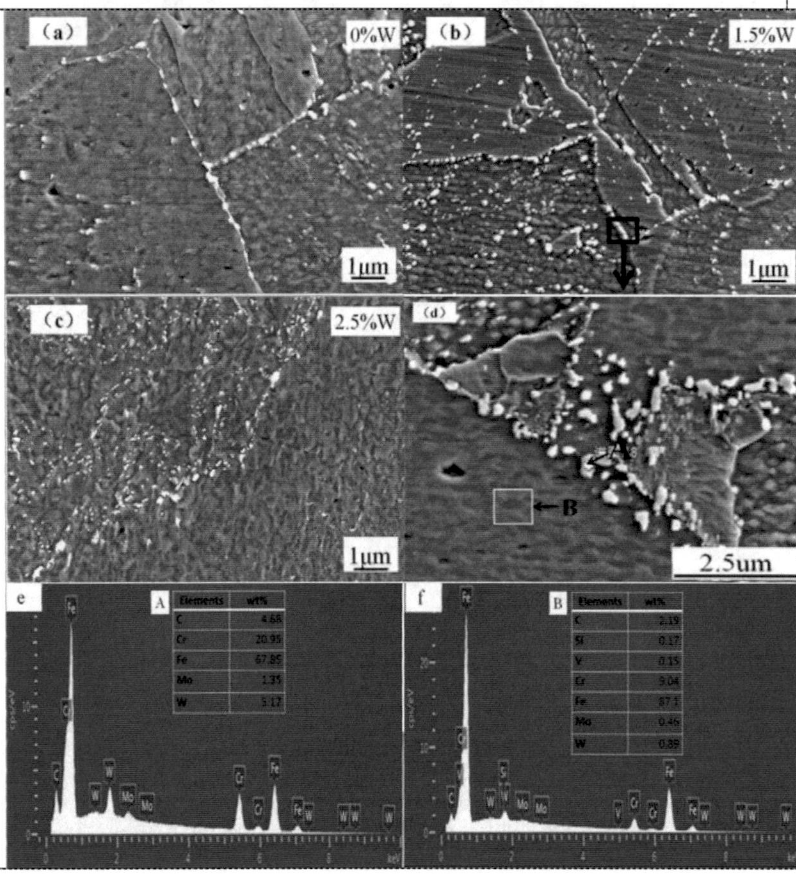

Fig. 4 Scanning electron microscope (SEM) images of 9%Cr heat-resistant steel at 650°C. (a) 9Cr-0W; (b) 9Cr-1.5W, (c) 9Cr-2.5W, (d) 9Cr-1.5W-partial enlarged drawing, and Energy Dispersive Spectrometry (EDS) (e) and (f)

Figure 3 displays the optical microstructure of tempered and normalized martensitic without aging containing different contents of W element ((a) 0W; (b) 1.5W, and (C) 2.5W). The matrix primarily consisted of lath martensite structure in prior austenite grain (PAG). With the increase in W content, the martensite lath and austenite grains were refined.

For the analysis of morphology and phase of the precipitates, the microstructure of the specimen was observed using Scanning electron microscope (SEM) images (see figure 4 (a-d)), Energy Dispersive Spectrometry (EDS) analysis (see figure 4(e, f))), and X-ray diffraction (XRD) measurements (see figure 5). The results demonstrated that the precipitates were $M_{23}C_6$ while no other types of precipitates were identified. The average particle size and density of carbides in the SEM images were measured using Image-Pro-Plus software. The results are presented in figure 6. As the concentration of W element increased, the number of $M_{23}C_6$ gradually increased and the size of $M_{23}C_6$ decreased; namely, the W element induced the refining of $M_{23}C_6$. Through EDS measurements, the primary components of $M_{23}C_6$ were Fe, Cr, C, Mo, and W, and the W element in the precipitate was much higher than that in the matrix. These results suggest that W was mainly enriched in $M_{23}C_6$ during normalizing and tempering.

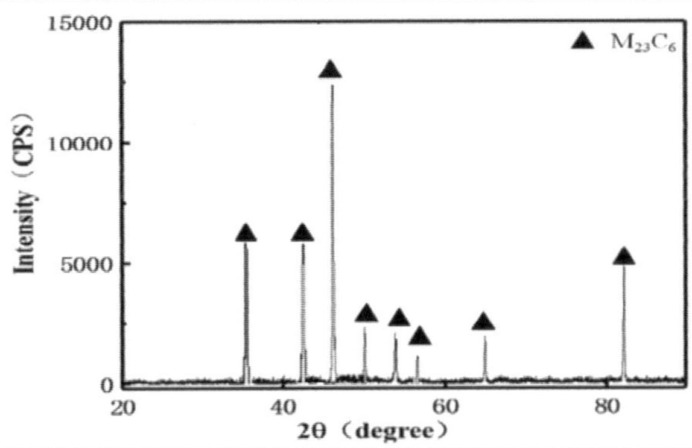

Fig. 5 X-ray diffraction (XRD) pattern of heat-treated specimen of 9 Cr1.5W at 650°C

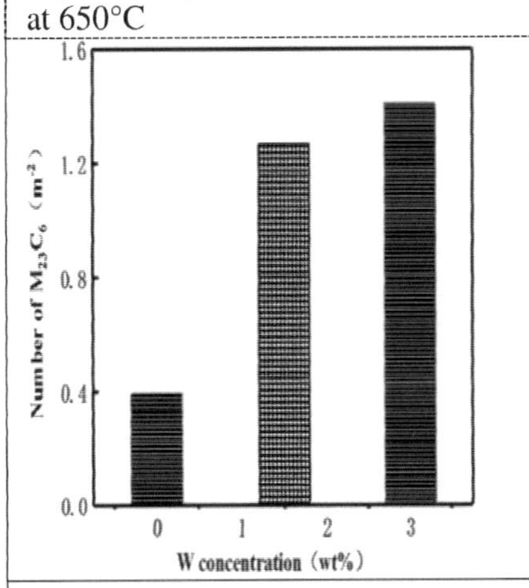

Fig.6 (a) Number of $M_{23}C_6$ as function of W concentration

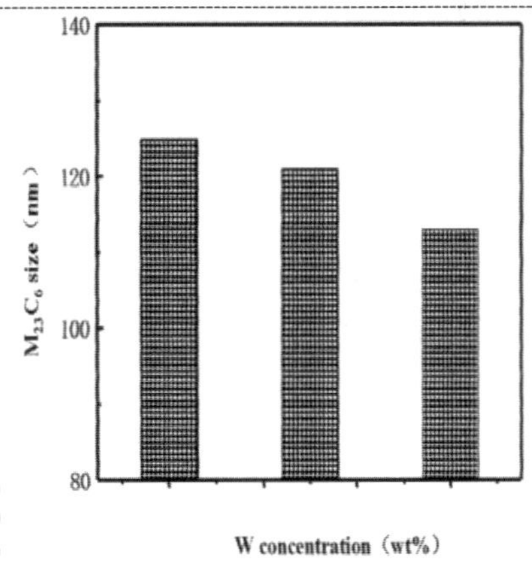

Fig.6 (b) Size of $M_{23}C_6$ as function of W concentration

As shown in figure 7 (a) and figure 8 (a), the creep curves at a high stress, 120 MPa, can be divided into three stages; the primary creep stage in which the creep strain decreases gradually with time, the secondary creep stage in which the creep strain basically stays unchanged, and the accelerated creep stage in which the creep strain increases rapidly with time until fracture. At the initial stage of creep of 1000 s, the creep rates of 9Cr steels with different W concentrations were approximately the same. For a longer period of time above 1000 s, with the increasing in W concentration, the minimum creep rate and elongation decreased while the time to fracture increased (see figure 9). These results are consistent with the previous conclusions [18]. For a low stress 30 MPa (see figure 7 (b) and 8 (b)), the creep curves included only primary creep stage. With the increasing in W concentration, strain rate gradually decreased. These results show that W can enhance the creep strength of 9Cr steel under high stress and low stress conditions but reduce the plasticity.

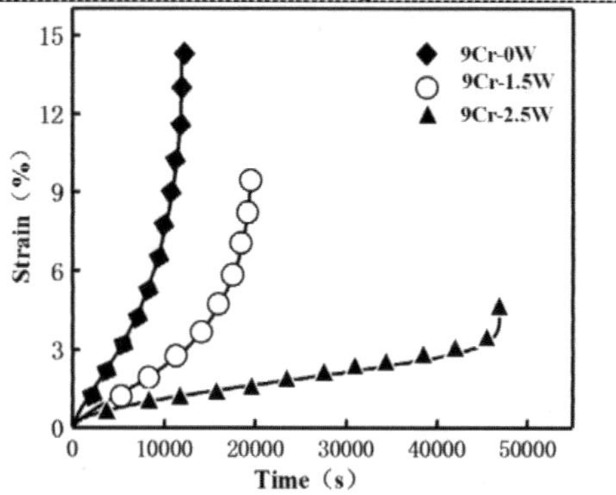

Fig. 7 (a) Creep curves of 9%Cr heat-resistant steel as the function of W concentration with different stress contents at 650°C and 120 MPa

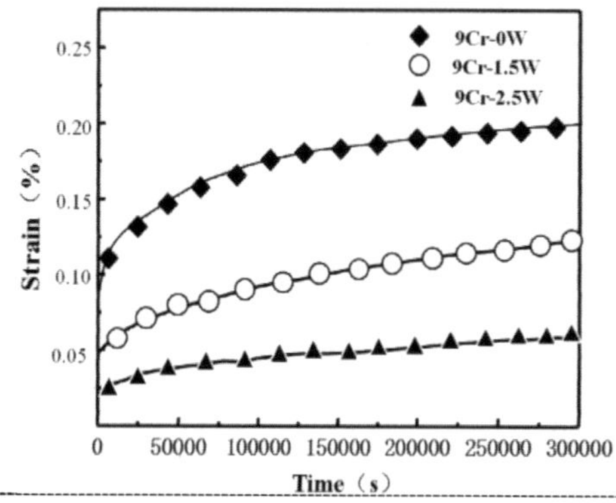

Fig. 7 (b) Creep curves of 9%Cr heat-resistant steel as the function of W concentration with different stress contents at 650°C and 30 MPa

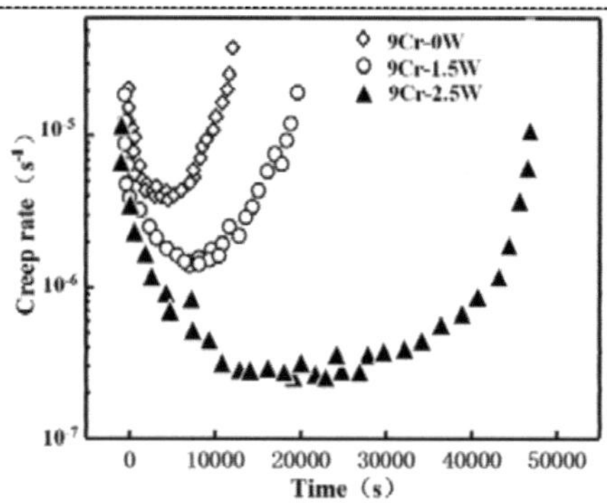

Fig. 8 (a) Creep rates in log scale versus time curves for 9Cr steels as the function of W concentration with different stress at 650°C and 120 MPa

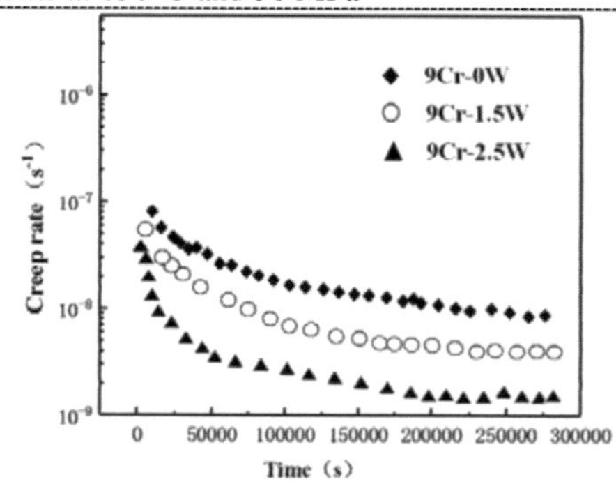

Fig. 8(b) Creep rates in log scale versus time curves for 9Cr steels as the function of W concentration with different stress at 650°C and 30 MPa.

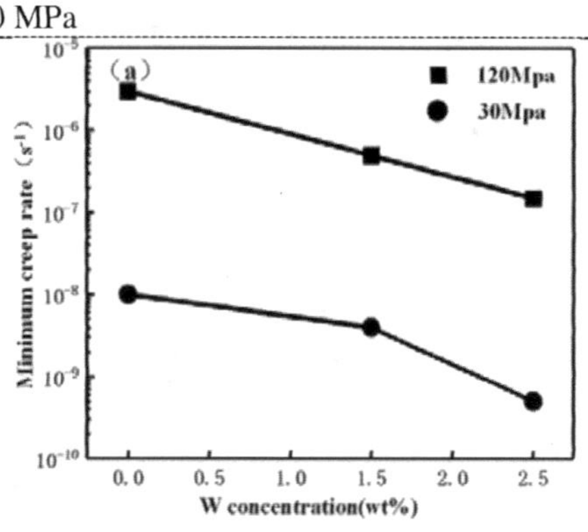

Fig. 9(a) Dependence of the minimum creep rate on W concentration in 9Cr steel creeping at 650°C, 120 MPa and 30 MPa,

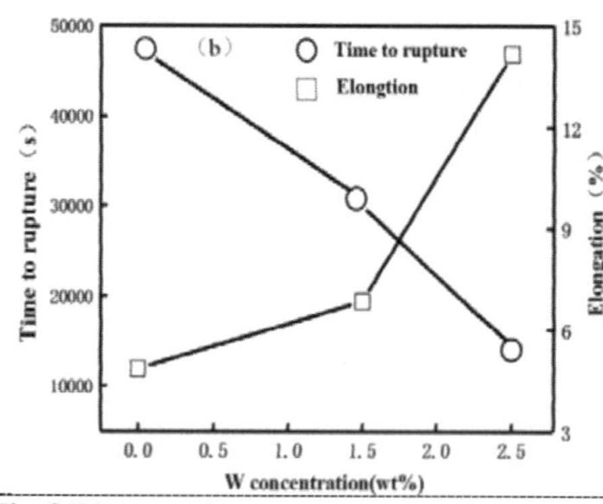

Fig.9(b) Dependence of elongation and time to rupture on W concentration in 9Cr steel creeping at 650°C, 120 MPa

Schematic drawings of the effects of W concentration on the creep properties are presented in figure 10. With the increase in W concentration, the concentration of W in both the matrix and $M_{23}C_6$ carbides increased, stabilizing the $M_{23}C_6$ carbides, which led to the refinement of $M_{23}C_6$ and prevention of the migration of lath boundary. The phenomenon provided a significant hindrance to the dislocation movement, leading to an improvement in the creep strength.

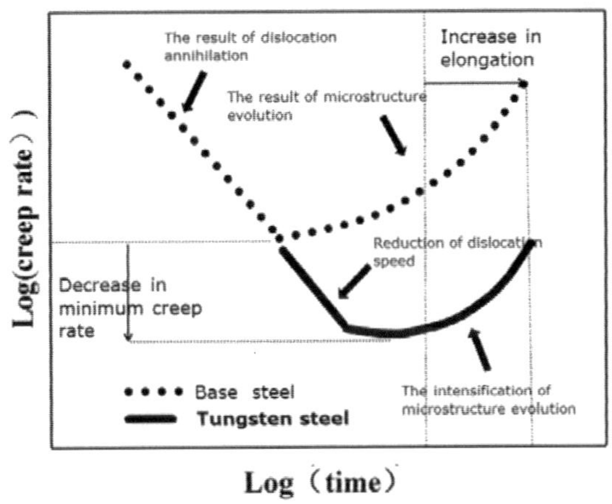

Fig. 10 Schematic drawings of the effects of W concentration on creep properties [18]

4. Conclusion

The microstructure and creep properties under different tungsten concentrations in 9%Cr ferritic heat-resistant steel were investigated to study the effect of W content on the creep properties. The primary results are summarized as follows:

(1) With the increase in tungsten concentration, the number of $M_{23}C_6$ increased while the particle size decreased.

(2) With the increasing in W concentration, the minimum creep rate and elongation decreased, while the time to fracture increased.

(3) The increase in W concentration led to the refinement of $M_{23}C_6$, which enhanced the creep strength, but reduced the creep plasticity.

5. References

[1] Aghajani A, Somsen C and Eggeler G 2009 On the effect of long-term creep on the microstructure of a 12% chromium tempered martensite ferritic steel *Acta Materialia*.**57** 5093-5106.

[2] Panait C G, Bendick W , Fuchsmann A, Gourgues-Lorenzon A F and Besson J 2010 Study of the microstructure of the grade 91 steel after more than 100000h of creep exposure at 600°C *International Journal of Pressure Vessels & Piping*.**87(6)** 326-335.

[3] Yagi K and Abe F 2001 Creep-resistant steels *Encyclopedia of Materials Science & Technology* .**16(9)** 1840-1845.

[4] Kaybyshev R O, Skorobogatykh V N and Shchenkova I 2010 New martensitic steels for fossil power plant *Physics of Metals & Metallography* .**109(2)** 186-200.

[5] KitaharaH , Ueji R , Tsuji N and Minamino Y 2006 Crystallographic features of lath martensite in low-carbon steel *Acta Materialia* .**54(5)** 1279-1288.

[6] Abe F 2009 Analysis of creep rates of tempered martensitic 9%cr steel based on microstructure evolution *Materials Science & Engineering A (Structural Materials: Properties, Microstructure and Processing)*.**510-511(none)** 64-69.

[7] Yuantao W, Mingjia W, Yan G, Tao C, Lei Z, Xuan M, Qian and Jing H 2015 Study on the nucleation and growth of laves phase in a 10% cr martensite ferritic steel after long-term aging *Journal of Alloys and Compounds* .**621** 93-98.
[8] Abe F, Araki H and Noda 1991 The effect of tungsten on dislocation recovery and precipitation behavior of low-activation martensitic 9cr steels *Metallurgical Transactions A (Physical Metallurgy and Materials, Science)* .**22(10)** 2225-35.
[9] Abe F 2005 Effect of fine precipitation and subsequent coarsening of fe2w laves phase on the creep deformation behavior of tempered martensitic 9cr-w steels *Metallurgical and Materials Transactions A (Physical Metallurgy and, Materials Science).* **36(2)** 321-332.
[10] Tkachev E, Belyakov A and Kaibyshev R 2018 Creep behavior and microstructural evolution of a 9%cr steel with high B and low N contents. *Materials Science and Engineering.* **725** 228-241.
[11] Kostka A, Tak K G , Hellmig R J , Estrin Y and Eggeler G 2007 On the contribution of carbides and micrograin boundaries to the creep strength of tempered martensite ferritic steels *Acta Materialia* .**55(2)** 539-550.
[12] Yang F, Hou J, Gao S , Wang C and Zhou L 2017 The effects of boron addition on the microstructure stability and mechanical properties of a ni-cr based superalloy *Materials Science and Engineering* .**715** 126-136.
[13] Horiuchi M T, Igarashi and Abe F 2002 Improved Utilization of Added B in 9Cr Heat-Resistant Steels Containing W *ISIJ International.* **42** S67-S71.
[14] Zheng Y, Wang F, Li C, Cheng J and Li Y 2018 Effect of compound addition of nb-b on hot ductility of cr-mo alloy steel *Materials Science and Engineering A.* **715** 194-204
[15] Qiuzhi G, Yanan Z , Hailian Z, Huijun L, Fu Q, Jian H, Cheng L, Bintao W, Yao L and Yan M. Possible 2001 Effect of Co on Coarsening of $M_{23}C_6$ Carbide and Orowan Stress in a 9% Cr Steel *ISIJ International*, **41** 356-360.
[16] Yu H, Xu W and Sybrand V D Z 2017 On the cobalt tungsten/chromium balance in martensitic creep resistant steels *steel research international* **89** 1700320.
[17] Abe F 2001 Creep rates and strengthening mechanisms in tungsten-strengthened 9cr steels *Materials Science & Engineering A (Structural Materials:, Properties, Microstructure and Processing).* **319-321** 770-773.
[18] Abe F and Nakazawa S 1992 The effect of tungsten on creep *Metallurgical Transactions A.***23(11)** 3025-3034.
[19] Han H G, Shen J J and Xie J X 2018 Effects of Precipitates Evolution on Low Stress Creep Properties in P92 Heat-resistant Steel *Scientific Reports*, **8.**
[20] Ishibashi M, Fujimoto K, Ken-Ichi I, Hata S and Nakashima H 2009 High Temperature Deformation Behavior of Sn-Based Solder lloys under Low Stress Conditions by a Helical Spring Creep Testing Method *J. Japan Inst.* **73** 373–380.
[21] Crossland I G , Jones R B and Lewthwaite G W 1973 The use of helically coiled springs in creep experiments with special reference to the case of bingham flow *Journal of Physics D (Applied Physics).***6(9)** 1040---46.
[22] Shen JJ 2013 Abnormal transient creep in fine-grained Al-5356 alloy observed at low strain rates by high-resolution strain measurement *Transactions of Nonferrous Metals Society of China*, **23** 1209-114.
[23] Zhao P S, Shen J J and Zhang H 2016 Short-term creep behavior in p91 heat-resistant steel at low stress*Materials Science Forum* **850** 922-926.

Acknowledgments
This project was supported by the National natural science foundation of China (Grant No.51605330), Natural science foundation of tianjin (No.18JCYBJC88700)

IOP Conf. Series: Materials Science and Engineering **838** (2020) 012004 doi:10.1088/1757-899X/838/1/012004

Numerical study on porosity distribution in casting under micro- amplitude vibration condition

K You, L Rao*, X Bai and T Y Miao

College of Mechanics and materials, Hohai University, Nanjing 211100, China

*Corresponding author's e-mail address: raol@hhu.edu.cn

Abstract. Combining with Pequet's porosity model, this paper coupled vibration effect into the heat transfer models based on the turbulent heat transfer theory, on this basis porosity distribution of alloys under micro-amplitude vibration conditions can be analyzed. The simulated porosity results of ZL201 alloy show that appropriate vibration frequency (10Hz~30Hz) promoted the mass feeding and capillary feeding by increasing the Dendrite Arm Spacing (SDAS) and heat transfer specific surface area of mass liquid channel, thus efficiently decreasing porosity distribution in casting. However, too high vibration intensity (40Hz~50Hz) may deteriorate porosity distribution in casting due to insufficient feeding capacity. The vibration model provides a theoretical basis for the study on effect of mechanical vibration on casting, which is significant for the optimization of vibration parameters and porosity prevention in casting technology.

1. Introduction

Porosity is one of the major defects in castings, which results in a decrease of the mechanical properties. Convection caused by vibration can shorten the solidification time of hot spot, break up dendrites and expand the feeding channel [1, 2], thus decreasing porosity in casting. But unreasonable vibration parameters may lead to serious segregation, hot cracks and so on.

Numerical method is convenient for investigating the solidified structure of alloys. However, few numerical studies combine mechanical vibration with porosity formation. The random distribution of nucleation and growth of voids should be considered during the solidification process. Ch. Pequet [3] et al developed a porosity model for micro-porosity, macro-porosity formation during the solidification of alloys using a mushy-zone refinement method, which superimposes fine and regular finite volume grids onto the finite-element mesh used for the heat-flow computations, and solves the governing equations of micro-porosity with appropriate boundary conditions. The porosity model has been widely used in shrinkage porosity simulation.

The influence of vibration on casting has been widely discussed [4, 5]. Turbulence caused by vibration can accelerate the heat transfer, which will affect the porosity distribution. This paper established a theoretical model of vibration casting connecting with Pequet's model, which is significant for porosity prediction and optimization of vibration parameters in vibration casting.

2. Mathematical models

2.1. Heat transfer equation

Under micro-vibration amplitude condition, the Reynolds number in the molten pool is very small, so the heat transfer in the molten pool is mainly in the way of heat conduction. The dominant equation is shown as:

$$\rho c_p \frac{\partial T}{\partial t} = \frac{\partial}{\partial x}\left(\lambda \frac{\partial T}{\partial x}\right) + \frac{\partial}{\partial y}\left(\lambda \frac{\partial T}{\partial y}\right) + \frac{\partial}{\partial z}\left(\lambda \frac{\partial T}{\partial z}\right) + Q_V . \quad (1)$$

Where ρ is the density; c_p is the specific heat; T is the temperature; t is the time; λ is the thermal conductivity; Q_V is the energy input by vibration. The convection of liquid melt can be reflected by increasing the thermal conductivity, which is also called effective thermal conductivity λ:

$$\lambda = \lambda_m + \lambda_t ; \quad \lambda_t = \frac{1}{P_{rt}} \rho C_p C_\mu \frac{\kappa^2}{\varepsilon} . \quad (2)$$

Where λ_m is the static thermal conductivity of the metal pool, λ_t is turbulent thermal conductivity generated by convection[6]; ε is the turbulent flow dissipation rate, κ is the pulsating kinetic energy. C_μ is the experimental constant and usually take 0.09 in engineering. For micro-vibration amplitude, liquid motion in molten pool is disorderly, to weaken the effect of vibration direction, three-dimensional vibration energy Q_{Vi} is exerted on the casting mould during solidification. The same frequency f and amplitude A are exerted at each orthogonal coordinate direction, and the difference of phase angle between each two coordinate axis is 120degrees.
For the vibration energy Q_V, it can be expressed as:

$$Q_V = 3Q_{Vi}; \quad Q_{Vi} = 8\pi^2 \rho A^2 f^3, \quad i = x, y \text{ and } z . \quad (3)$$

In semi-solid metals, a solid-phase network is formed when the solid fraction g_s exceeds 0.4, which interrupts the liquid phase [7, 8], thus $Q_V = 0$ and $\lambda_t = 0$ in this condition.

2.2. Porosity nucleation and growth model

2.2.1. Mass conservation with porosity

The mass conservation is a continuity equation as follow:

$$div(\rho_l g_l v_l) - \rho_l \frac{\partial g_p}{\partial t} = -\frac{\partial \langle \rho \rangle}{\partial T} \frac{\partial T}{\partial t} . \quad (4)$$

where $\langle \rho \rangle = \left(\rho_s g_s + \rho_l (1 - g_s)\right)$ and $g_l = 1 - g_s - g_p$, ρ_s is the specific mass of the solid, g_s is the solid fraction, ρ_l is the density of the liquid, and v_l is the velocity of the liquid in solid skeleton.

2.2.2. Porosity nucleation criterion

In the absence of gas dissolved in the liquid metal: pore will nucleate if decreasing liquid pressure reaches minus of the overpressure due to capillarity effect:

$$p_{nuc} = -\Delta p_r; \quad \Delta p_r = \frac{2\sigma_{lg}}{r} . \quad (5)$$

Where Δp_r is the overpressure of nucleation; σ_{lg} is the interfacial tension between the liquid and the pore, and r is the radius of curvature of the pore.

2.2.3. Porosity growth model

Three cases are considered as follow (Figue.1 (b)):

(1) Case 1: A region where solid fraction is lower than the "mass feeding limit": liquid contraction and solidification shrinkage are compensated by liquid from the top of critical region.

(2) Case 2: A region is higher than “mass feeding limit”, but lower than the inter-dendrite "gravity draining limit": Liquid contraction and solidification shrinkage are compensated by liquid with solid fraction lower than “gravity draining limit”, which can be solved by Darcy's flow:

$$g_l v_l = -\frac{K}{\mu}(\nabla p_l - \rho_l g);\; K(g_s,\lambda_2) = \frac{(1-g_s)^3}{g_s^2}\frac{\lambda_2^{\,2}}{180}. \quad (6)$$

Where p_l is local pressure in liquid; K is the permeability of the solid skeleton; λ_2 is the secondary dendrite arm spacing; v_l is the dynamic viscosity and g is the gravity vector.

(3) Case 3: Solid fraction is above "gravity draining limit": liquid contraction and solidification shrinkage are compensated by distributed inter-dendrite micro-shrinkage increment.

2.3. Model and calculation of parameters

Three-dimensional finite element model with a cylindrical shape has been established ((figue.1(a))). Standard ZL201 aluminum alloy is used. The mould material is H13 and the insulation material is High alumina brick. The pouring temperature is 710℃. Both the initial temperature of mould and insulation is 20℃. The interfacial heat transfer coefficients of mould/ingot, insulation/ingot and mould/insulation are 2300, 500 and 1000 W/m^2-K, respectively.

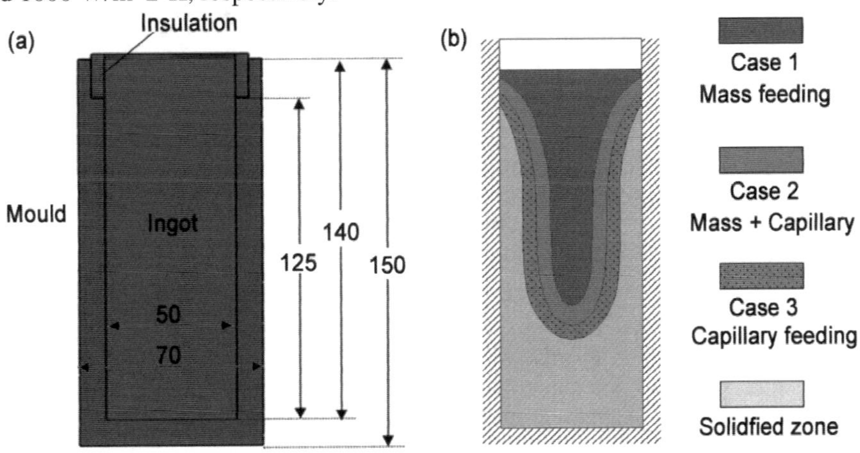

Figure 1. The casting system (a) and three cases of porosity model (b).

3. Results

3.1. Solidification distribution

In this section, casting without vibration and casting with vibration frequencies from 10Hz to 50Hz were simulated. It can be seen from figue.2 (upper part) that lower temperature gradient formed in all castings with vibration than that of casting without vibration, which is the same as results reported in reference [9]. Here the solid fraction in scale of 0~40% is supposed to be lower than the "mass feeding limit" and solid fraction in scale of 40~60% is supposed to be higher than the "mass feeding limit" but lower than inter-dendrite "gravity draining limit".

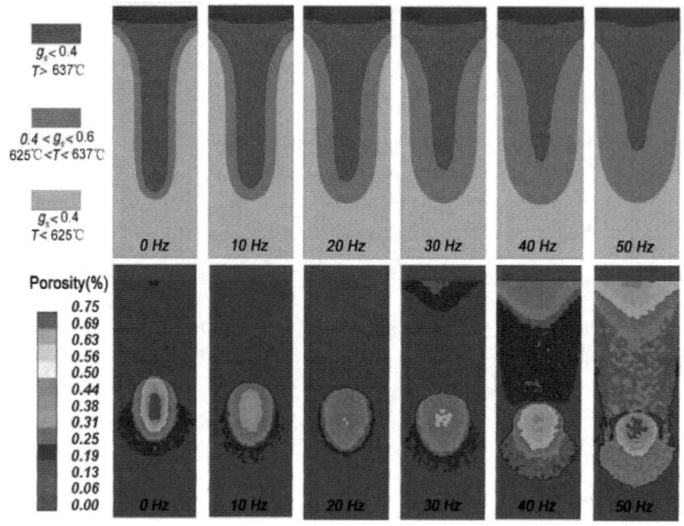

Figure 2. Solidification at 60% of solid fraction and final porosity distribution.

It can be seen that the depth of mass liquid feeding (0~40%) channel in all castings with vibration were shallower than that in casting without vibration at the same solid fraction, and the depth of mass liquid feeding became shallower gradually with increasing vibration frequency. The area of capillary feeding channel (40~60%) in all castings with vibration is broader than that of casting without vibration, and the area of capillary feeding channel became boarder gradually with the increasing frequency.

3.2. Porosity distribution

Figue.2 (lower part) shows porosity distribution of all castings with and without vibration, it can be seen porosity was mainly located at the top and central part of castings. Because the top position is the last solidification position, thus only small ratio of porosity was formed. Gradually decreasing porosity ratio and area was formed for castings with vibration when the frequency increasing from 10Hz to 30Hz, showing better casting quality than casting without vibration. However, bigger porosity ratio and area was formed both at the top and the central part for castings with vibration frequency of 40Hz and 50Hz than that of casting without vibration, which indicated inadequate feeding ability. This trend is the same as the conclusion of some experimental results in reference [10, 11].

4. Discussion

Porosity is mainly affected by combination effects of mass feeding and capillary feeding. Figue.3 (a) shows the aspect of mass feeding channel of all castings. Higher aspect symbols lower specific surface area of mass feeding channel. With the increasing vibration frequency from 10Hz to 30 Hz, the mass liquid channel became broader, which promoted compensation of mass feeding. But when the vibration frequency was above 40 Hz, porosity distribution area became enlarged, which can't be explained only by the aspect of mass feeding channel.

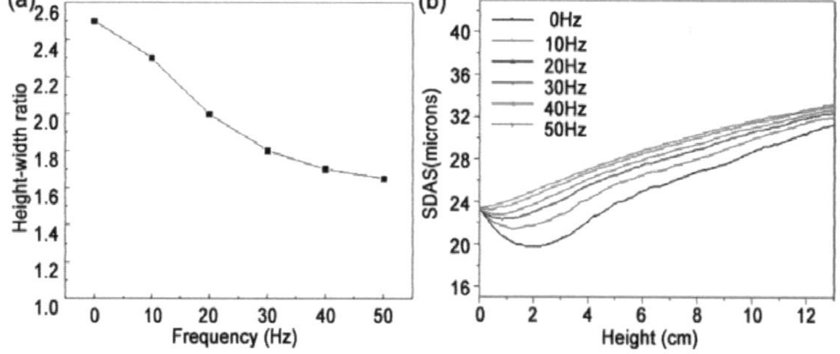

Figure 3. Length-width ration of mass feeding zone (a) and SDAS of capillary feeding (b).

According to Darcy's Flow, larger λ_2 may symbols stronger capillary feeding ability. Figue.3 (b) shows the λ_2 at the center line of all castings, it can be seen increasing λ_2 were formed in castings with vibration than casting without vibration, indicating higher capillary feeding ability in castings with vibration (10~30Hz) than casting without vibration. However, when the solid fraction of 40~60% is too board (40Hz and 50Hz), Mush zone may solidify completely before capillary feeding is achieved (or not adequate feeding liquid). Thus, bigger porosity area is easily accumulated in castings with frequency of 40Hz and 50Hz than that of casting without vibration. In addition, high vibration frequency increases the internal energy of liquid melt, which slows down the cooling rates of solidification front and increases porosity ratio in casting.

5. Conclusion

Based on turbulent heat transfer theory, this paper inserts the effect of mechanical vibration into heat transfer model, on this basis porosity distribution of alloys under micro-amplitude vibration condition can be analyzed combining Pequet's refined porosity model. The simulated results of ZL201 alloy showed that enhancing vibration frequency from 10Hz to 30Hz increasingly promoted the mass gravity draining feeding and capillary feeding in casting, thus reducing the porosity distribution. However, when vibration frequency was above 40Hz, porosity distribution in casting was deteriorated due to the increase of capillary complementary area and vibration energy. The vibration model provides a theoretical basis for predicting porosity distribution in casting under vibration conditions and is significant for the improvement of vibration parameters in casting technology.

Reference

[1] Javid F, Wang P, Shanian A, et al. Architected Materials with Ultra-Low Porosity for Vibration Control[J]. *Advanced Materials* 2016.

[2] Yan H, Rao, Yuansheng, He, Ru. Morphological evolution of semi-solid Mg2Si/AM60 magnesium matrix composite produced by ultrasonic vibration process[J]. *Journal of Materials Processing Tech* 2014 214(3) 612-619.

[3] Pequet C, Rappaz M, Gremaud M. Modeling of microporosity, macroporosity, and pipe-shrinkage formation during the solidification of alloys using a mushy-zone refinement method: Applications to aluminum alloys[J]. *Metallurgical and Materials Transactions* A (Physical Metallurgy and, Materials Science), 2002, 33(7) 2095-06.

[4] Fand R M, Peebles E M. A Comparison of the Influence of Mechanical and Acoustical Vibrations on Free Convections from a Horizontal Cylinder [J]. *Journal of Heat Transfer* 1961, 84(3) 39.

[5] Gencalp S, N. Saklakoglu. Effects of Low-Frequency Mechanical Vibration and Casting Temperatures on Microstructure of Semisolid AlSi8Cu3Fe Alloy[J]. *Arabian Journal for Science & Engineering* 2012, 37(8) 2255-67.

[6] Macro-and Microstructure Evolution of 5CrNiMo Steel Ingots during Electroslag Remelting Process[J]. *Journal of Iron and Steel Research, International* 2014, 21(7) 644-652.

[7] Lashkari O, Ghomashchi R . The implication of rheology in semi-solid metal processes: An overview[J]. *Journal of Materials Processing Technology* 2007, 182(1-3) 229-240.

[8] Kirkwood, D. H. Semisolid metal processing[J]. *International Materials Reviews* 1994, 39(5) 173-189.

[9] Wei B. Unidirectional dendritic solidification under longitudinal resonant vibration [J]. *Acta Metallurgica Et Materialia* 1992, 40(10) 2739-51.

[10] Kocatepe K. Effect of low frequency vibration on porosity of LM25 and LM6 alloys[J]. *Materials & Design* 2007, 28(6) 1767-75.

[11] Shukla D P, Goel D B, Pandey P C. Effect of vibration on the formation of porosity in aluminium alloy ingots [J]. *Metallurgical Transactions B* 1980, 11(1) 166-168.

Acknowledgments

We are grateful for grants from the National Natural Science Foundation of China (No. 51775167), and the Qing Lan Project of Jiangsu province.

IOP Conf. Series: Materials Science and Engineering **838** (2020) 012005 doi:10.1088/1757-899X/838/1/012005

Evaluation of dew point corrosion in acid atmosphere by wire beam electrode

Bo Zhao[1,a],Yuxin YU[1,b],Jing GUO[1,c], Tianyu ZHOU[1,d], Shiwen ZOU[2,e] and Tong XU[1,f]

[1]Materials Research and Evaluation Center, China Special Equipment Inspection and Research Institute, Beijing, 100013, China
[2]Aerospace Research Institute of Materials & Processing Technology, Beijing, 100076, China

E-mails: azhaobo19840626@163.com,byuyuxin@csei.org.cn,cguojing@csei.org.cn, dzhoutianyu@csei.org.cn, ezoushiwen908@163.com, fxutong@csei.org.cn

Abstract. In this article, wire beam electrode (WBE) was used to evaluate the corrosion behavior of ND steel in environmental acid atmosphere with different partial pressure of CO_2. Meanwhile, corrosion products and surface morphology analysis also used to support this research. The results showed that the corrosion behavior began from the edge of droplet in dew point corrosion, and gradually spread to the center of it. The spread speed would increasing with CO_2 partial pressure enhance, which was 24h in 5% CO_2 and 4h in 50% CO_2. Corrosion current density in the edge of droplet can form the "cathode-anode-cathode ring" structure, and disappears gradually as the corrosion time was going. Corrosion morphology observation results showed three ring shapes region and different elemental composition of different corrosion products, which is correspondence with the "cathode-anodWe-cathode ring" structure measured in WBE experiments. The results showed that the reaction gradually transferred to the uniform corrosion on electrode surface when the dew point corrosion reaction reaching the late stage. It comes from the dissolution, diffusion and reaction of gaseous corrosion medium of CO_2 and O_2.

Keywords: Dew point corrosion, Wire beam electrode, Acid atmosphere

1. Introduction

Dew point corrosion (DPC) is the most important type of high temperature equipment corrosion[1-2]. Some researchers' consideration that DPC is a kind of special corrosion in thin electrolyte film, and the corrosion rate was due to the rate of corrosion media mass-transfer process. As for DPC, the corrosion media is different to the common thin electrolyte film, the acid atmosphere such as SO_2, SO_3, CO_2 and HCl are became the corrosion media instead O_2 to attack the base metal[3-4].

2. Experimental

2.1. Materials preparation

The metal we chose is 09CrCuSb steel (ND steel), which is a new type of low alloy steel used in DPC environment. The chemical compositions showed TAB 1, referring to standard of GB/T150.2-2015. As the result showed, the chemical compositions of this steel are kept according to the requirements of the standards, and it used to make the electrodes and observation samples.

Table 1. The chemical compositions of ND steel.

Ele.	C	Si	Mn	P	S	Cr	Cu	Sb
Std.	≤0.12	0.20~0.40	0.35~0.65	≤0.030	≤0.020	0.70~1.10	0.25~0.45	0.04~0.10
Samp.	0.10	0.16	0.42	0.010	0.008	1.05	0.33	0.08

Fig.1 is the metallographic structure of ND steel. As the results, its microstructure is ferrite with a few pearlite and carbides.

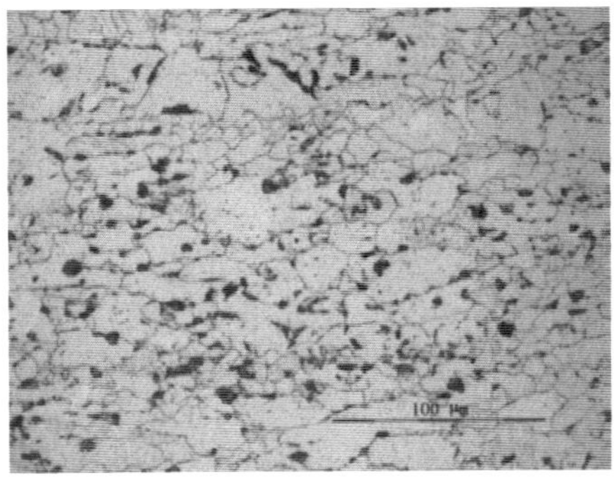

Figure 1. The metallographic structure of ND steel.

2.2. Wire beam electrode measurement

In order to study the transfer mass-transfer characteristics, wire beam electrode (WBE) measurement were used to evaluate the surface current distribution on the ND electrode. The WBE is formed with an11×11 wire bundle electrode arrays (121 electrodes). Every electrode cuts into a cylinder with 1mm surface diameter, and sealed into an epoxy resin cube with 0.3mm interval. The working surface of the WBE was subsequently ground with 150 to 2000 grit emery papers, and then cleans it with distilled water and acetone after polishing. The measurement was performed with a Corttest CT520 WBE scan workstation with the surface current scanning model, and the result presents current distribution on surface.

We used a work-holding device to hold the WBE work surface keep up, and dropped 0.5ml distilled water on it formed a single corrosive droplet. In order to prevent the droplet evaporation, WBE set in a constant temperature and constant humidity chamber with 25°Cand 100%RH. The internal dimensions of the chamber ($l\times w\times h$) is 100×100×50mm, and inject 5%and 50% CO_2 acid gas mixed by CO_2 and N_2 with flow rate of 100ml/min, making the saturated acid atmosphere. In this experiment, 121 electrodes which coupled together are disconnected successively, and the couples current between each electrode and the other electrodes were measured respectively. Scanning time are 0, 2, 4, 8, 24, 48 hours after corrosion occurrence, totally 6 times.

3. Results and discussion

3.1. Results of wire beam electrode measurement

Fig. 2 showed the change of current distribution on ND steel surface in5% CO_2 acid atmosphere. In this experiment, all surface of WBE cell was covered by distilled water droplets, and the abscissa and ordinate coordinates represent different electrode positions. It can be seen that the positions located on the edge and the center of WBE surface showed the lowest distribution current value at 0 h, and it presents those positions became cathodes in this reaction. At the same time, the distribution current close to the edge showed the highest distribution current value, and became anode. The corrosion current formed a structure of "cathode-anode-cathode ring" from the outer edge to the center under droplets, and the position of anode might be the area that corrosion reaction is the most serious. Therefore, the corrosion current reached 0.0265μAat 0h in anode area.

IOP Conf. Series: Materials Science and Engineering **838** (2020) 012005 doi:10.1088/1757-899X/838/1/012005

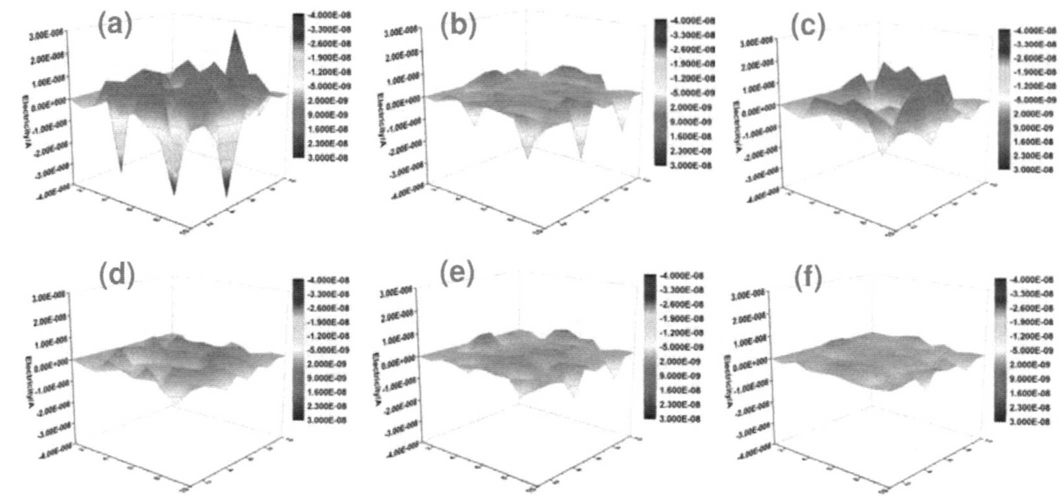

Figure 2. The change of current distribution on ND steel surface in5% CO2 acid atmosphere (a)0h, (b)2h, (c)4h, (d)8h, (e)24h, (a)48h.

Cathode current at the edge of WBE disappeared gradually as the corrosion time was going, and the position of the anode was not changed basically, but the current value decreases gradually. The maximum anodic current values reached 0.0157, 0.0066 and 0.0059μA at time after corrosion time of 2h, 4h, 8h respectively, which were 41%, 75% and 76% lower than 0h.The corrosion reaction tended to stable at 8h, and the electrode surface could not observe the characteristics of separation of anode and cathode. The corrosion proceeded more smoothly on the surface of the electrode, and the corrosion characteristics were mainly uniform corrosion.

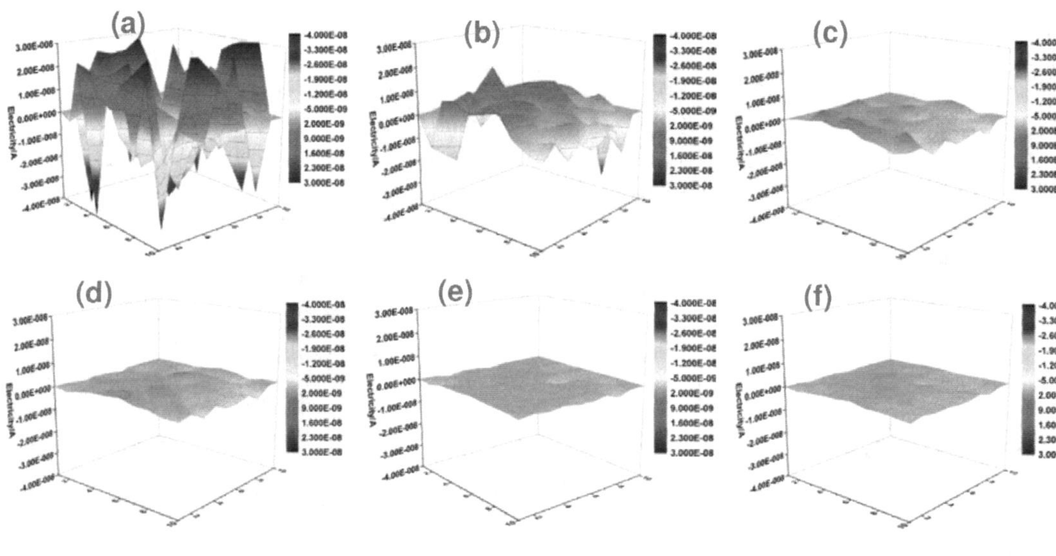

Figure 3. The change of current distribution on ND steel surface in50% CO2 acid atmosphere (a)0h, (b)2h, (c)4h, (d)8h, (e)24h, (a)48h.

Fig.3 showed the change of current distribution on ND steel surface in5% CO_2 acid atmosphere. As the results, the distribution of cathode and anode on WBE surface in 50% CO_2 environment is similar to that in 5% CO_2 environment at the initial stage of corrosion reaction, but it is much more severe. The maximum anodic current reached about 0.0328μA at 0 h, which is about 1.24 times higher than that in 5% CO_2 environment.

The current density of the anode decreases gradually as the corrosion time was going, and it is faster than that in 5% CO_2 environment. The separation of anode and cathode has been slight in electrochemical reaction at 4h, and then stabilized at 24h.That means that the corrosion degree on the WBE surface has declined, and it tended to the uniform corrosion.

Fig.4 showed the comparison of the maximum current on WBE surface with 5% and 50% CO_2. It can be seen that the maximum surface current of ND steel has similar change tendency in the two different concentration of CO_2content environment. All of them have the highest surface current at 0 h, which decreases rapidly as the corrosion time was going. 5% CO_2 reached steady state at 8h, while0% CO_2 reached steady state at 24h.

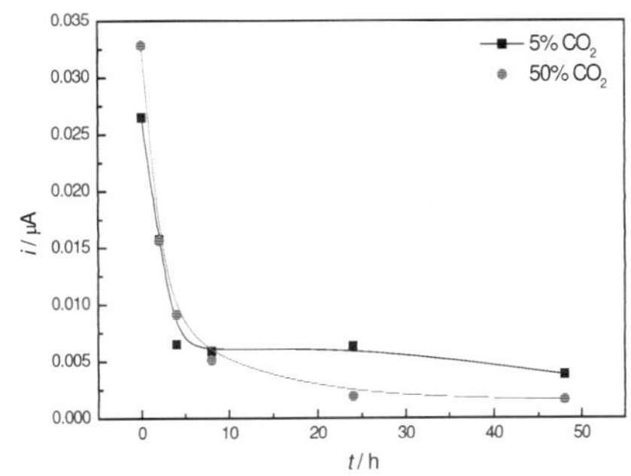

Figure 4. Comparison of the maximum current on WBE surface with5% and 50% CO_2.

The experiment results in 5% and 50% CO_2atmosphere all point to that the reaction gradually transferred to the uniform corrosion in all beam on the surface of every single electrodes as the corrosion time was going, and the characteristics of the galvanic reaction of the electrode decreased gradually.

4. Discussion

According to previous research, the corrosion reaction will be greatly different with different shape under thin electrolyte film. As for this research, the work electrode is ND steel, and the anodic reaction is Fe dissolution. The chemical equations is shown as follows[5-6]:

Anodic:

$$Fe \rightarrow Fe^{2+} + 2e^{-} \quad (1)$$

In the reaction process, the origin solution is distilled water with high resistivity, which can form the acidic solution while the vapor phase CO_2 dissolves in it. And then the solute can be ionized, with a part of depolarization ions change. The chemical equations are shown as follows:

Dissolution and Ionization of CO_2:

$$CO_2 + H_2O \leftrightarrow H_2CO_3 \quad (2)$$

$$H_2CO_3 \leftrightarrow H^+ + HCO_3^- \quad (3)$$

$$HCO_3^- \leftrightarrow H^+ + CO_3^{2-} \quad (4)$$

According to the equations, the ionization of CO_2 dissolved in water can be divided to two steps. The primary ionization constant $Ka(H_2CO_3)$ is 4.3×10^{-7} and the secondary ionization constant $Ka(HCO_3^-)$ is 5.61×10^{-11}in the room temperature of 20°C.Considering that saturated solubility of CO_2 is 0.04 mol/L in this situation, the theoretical maximum pH of this solution is pH=3.94.Therefore, both two types cathodic reaction of the electrodes of hydrogen evolution and oxygen absorption may exist at the same time. So that, the cathodic reactions are much more complex relatively, and the chemical equations is shown as follows:

Cathodic reactions:

$$O_2 + 4H_2O + 4e^- \rightarrow 4OH^- \quad (5)$$
$$2H^+ + 2e^- \rightarrow H_2 \quad (6)$$

According to the anodic and cathodic reactions, the corrosion process is shown in Fig.5. For the N_2 and CO_2 and moisture atmosphere continuous injection, it corresponds to a variety of atmospheric environments with N_2, CO_2 and O_2 in the experimental environment.N_2 and O_2 dissolve intoH_2O is process, while CO_2 dissolution is both physical and chemical process. So that, CO_2 reaching the solution interior is much more easily than O_2.

In the reaction process, the edge of droplet is the thinnest area, and forms a thin electrolyte film because of its shape. It is more easily that oxygen permeates into the droplet and reaches the surface of the substrate. So the edge of droplet is oxygen-rich and presents cathode characteristic[7]. The area beside the edge of droplet has low oxygen content and high CO_2 content, so it presents anode characteristic. At the same time, the center of electrode has a relatively thick electrolyte film, and solution is origin without CO_2 and its reaction. So that it also presents cathode characteristic, explaining why it can form the structure of "cathode-anode-cathode ring".

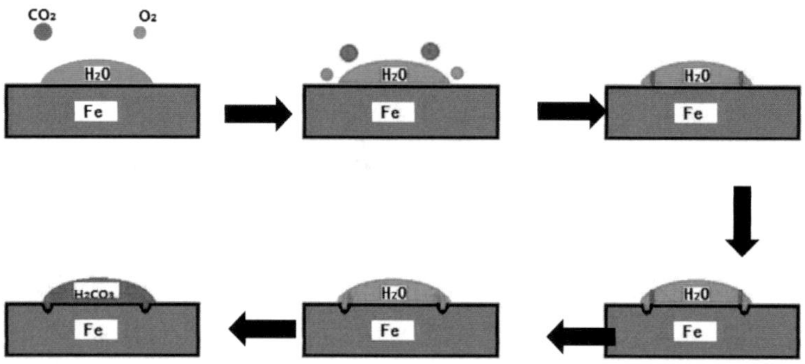

Figure 5. The mechanism of dew point corrosion of ND steel in acid atmosphere.

CO_2 dissolved gradual increase as corrosion reaction going, and concentration diffusion dominates the corrosion reaction process. H_2CO_3 diffuse into droplet interior, and H^+ will become the dominant anion in depolarization reaction. Corrosion scale of $FeCO_3$blocks on the electrode surface of in large quantities as corrosion reaction going, and hinders the mass transfer process of O_2 further. Therefore, the corrosion on the surface of the electrode becomes independent reaction in the later stage, and the difference between the cathode and anode regions is not obvious.

5. Conclusions

The dew point corrosion reaction under a droplet in acid atmosphere can form the "cathode-anode-cathode ring" structure from the outer edge to the center under droplets, and the anode region corrosive more serious. Cathode current at the edge of WBE disappear gradually as the corrosion time was going, but the time of reaching stationary current density on electrode surface in 50% CO_2 is more rapid than it in 5% CO_2 atmosphere.

The results point to that the reaction gradually transferred to the uniform corrosion on electrode surface when the dew point corrosion reaction reaching the late stage, and the characteristics of the galvanic reaction of the electrode decreased gradually. It comes from the dissolution, diffusion and reaction of gaseous corrosion medium of CO_2 and O_2.

6. Reference

[1] Dan Z H, et al. Applicability of constant dew point corrosion tests for evaluating atmospheric corrosion of aluminium alloys. *Corrosion Science* 53.5(2011):0-2014.
[2] Wang Y G et al Mechanism research on coupling effect between dew point corrosion and ash deposition." *Applied Thermal Engineering* 54.1(2013):102-110.
[3] Shu Y H, Wang F and Wu W Corrosion Behavior of Ti60 Alloy Coated with a Solid NaCl Deposit in O2 Plus Water Vapor at 500-700°C. *Oxidation of Metals* 52.5(1999):463-473.
[4] Huijbregts, W M M and R G I Leferink Latest advances in the understanding of acid dewpoint corrosion: corrosion and stress corrosion cracking in combustion gas condensates. *Anti-Corrosion Methods and Materials* 51.3(2004):173-188.
[5] Wilson and Robert E The Mechanism of the Corrosion of Iron. *Ind. eng. chem* 325.1(2002):91-6.
[6] Martinez, Sanja, Štern and Ivica. Inhibitory mechanism of low-carbon steel corrosion by mimosa tannin in sulphuric acid solutions. *Journal of Applied Electrochemistry* (2001).
[7] Shu, Y H, F Wang and W Wu. Corrosion Behavior of Pure Cr with a Solid NaCl Deposit in O2 Plus Water Vapor. *Oxidation of Metals* 54.5-6(2000):457-471.

Acknowledgement

The authors thank the China Special Equipment Inspection and Research Institute (CSEI) for the financial support through projects of Youth Foundation No. 2018-11, and General Administration of Quality Supervision, Inspection and Quarantine (AQSIQ) for the financial support through projects 2016QK209.

Effect of different microstructure on dynamical stab-resistant property of TA15 alloy

Wang Ruining[1], Sun Ruijun[1] and Yu Lingjie[1]

1School of Textile and Engineering, Xi'an Polytechnic University, Xi'an 710048, China

E-mail: wangruining123@163.com, bsunrunjun2002@163.com, clingjie.yu@xpu.edu.cn

Abstract. The effects of different microstructure on the dynamical stab-resistant property of TA15 alloy are investigated. The specimens were tested on the stab apparatus after heat treatment at temperature of 850, 930 and 1030°C, time of 30min and air cooled. The impact force-time curves, microstructure and fracture morphology of specimens were analyzed. The results showed that the properties of higher strength and plasticity are benefit to the performance of stab resistance of TA15 specimens, and the curves of 930°C and 850°C exhibit two peaks and longer time ranges, but the curve of 1030°C exhibits the shortest time to reach the peak of impact force, and the specimens heat treated at 850°C exhibit the better performance against the knife impactor because of fine grain and stronger grain boundary binding force to absorb the impact force energy.

Keywords: TA15 alloy, stab-resistant, microstructure, impact force

1. Introduction

For better performance of impact resistance property, a sandwich structure which consists of metal, fiber, and other soft materials [1] has been used as the energy absorption component in protection system in recent years. It has the property of high specific stiffness, low specific mass and high energy absorption. In the sandwich structure, the correct choice of metal plate is very important to the stab-resistant property with weight reduction and protection level improvement.

There are a lot of research on sandwich structure which using ceramic layer[2-3], aluminum layer[4-5] and polyethylene reinforced by Waterborne Polyurethane Resin or polymeric resin [6-7] to enhance the strength of the composite structure. According to the past several years research on sandwich structure, It can be known that high strength and lightweight properties is important to the stab-resistant performance. But there are few researches are available to study the stab-resistant performance of titanium which used as the strength and stiffness layer of the sandwich structure.

Titanium alloy has widely applied to many fields because of its excellent mechanical properties, low density and high specific strength [8-9]. It can be used as the high specific strength and stiffness layer to stop penetration by sharp point and cutting by blade edge effectively. TA15 (Ti-6.5Al-2Zr-1Mo-1V) alloy has been used as load-bearing structure in aerospace industry due to their excellent mechanical and physical properties [10-11]. In this research, The effect of different microstructure on stab resisting properties and energy absorption of TA15 titanium alloy was tested and analyzed.

2. Materials and experimental methods

2.1. Materials

The chemical composition of TA15 titanium alloy used in this work is Ti-6.5Al-2Zr-1Mo-1V(wt%). The testing specimens from TA15 sheet with 0.5mm thick were machined into square with 25mm in width and length for stabbing test. The initial microstructure of the samples consists of about 85% α phase and a few of β phase, as shown in Fig.1. In order to understand the effect of different microstructure on stab-resistant performance of TA15 alloy, the specimens were heat treatment at temperature of 850, 930 and 1030°C, time of 30min and air cooled to provide different microstructure.

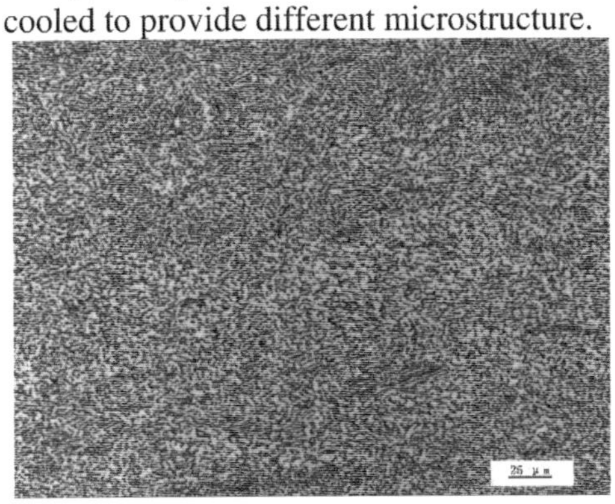

Figure 1. The initial microstructure of TA15 alloy.

2.2. *Experimental apparatus*

The stab-resistant properties of TA15 alloy with different microstructure were tested on the apparatus that drops freely under its own weight to strike the targets at a specified energy, which consists of an engineered knife blade on the drop mass. In the test, TA15 alloy specimen is 16 cm×16 cm in size, and the data of energy and displacement during the process of stab was recorded by the sensors connected with computer. The apparatus are shown in Fig.2.

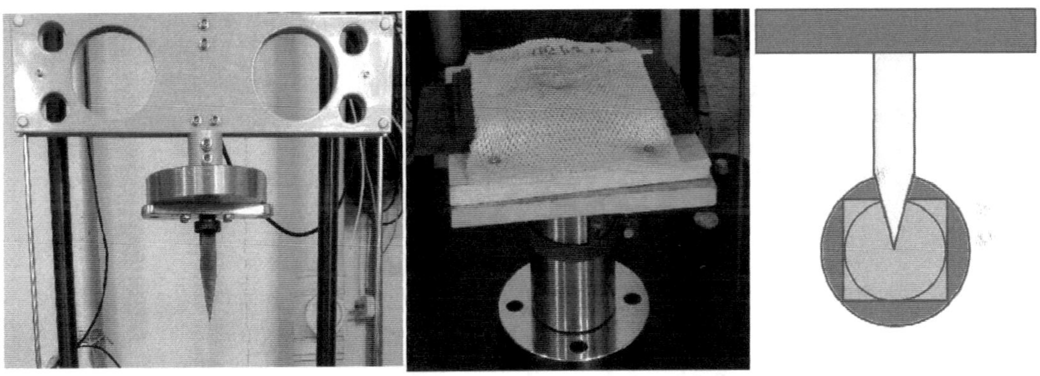

Figure 2. Schematic and instrument for dynamic stab test.

The tests specimens were divided into 2 groups according to the stab way. Group 1 was stabbed by knife directly on the circular jaws of the instrument, and group 2 was stabbed by knife with placing specimens on the rubber which is a simulation of human body. The circular jaws having inner and outer diameters of 72 mm and 105 mm, respectively. The apparatus that drops freely under its own weight to strike the specimens at a specified energy. The impactor is S1 knife based on NIJ Standard NIJ Standard–0115.00[12]as shown in Fig.2. The knife impacted the titanium sheet specimens at a speed of free falling movement, and the drop height was set 0.52 m for each specimen and the weight of knife was 2.33 kg. After test of stab, the microstructure and the crack section of TA15 alloy were observed by OM and SEM and the energy data were analyzed.

3. Results and discussion

3.1. *Relationship between stab-resistance and Mechanical properties of TA15 alloy*

For titanium alloy, there are more research about the use of aerospace, but other use of which is seldom to be observed. In order to understand the relationship among the property of stab-resistance, strength and plasticity, the mechanical properties of 0.5 mm TA15 alloy sheet with different heat treatment are shown in Fig.3. It can be seen that with the increasing of heat treatment temperature, the plasticity of TA15 alloy is decreasing, the strength is reduced apparently at 1030℃.

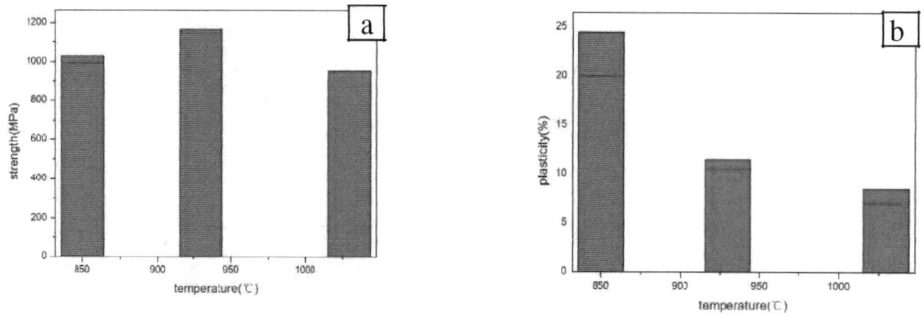

Figure 3. Mechanical properties of TA15 alloy with different heat treatment (a) strength (b) plasticity.

The drop tower stab performance of the TA15 alloy targets against the knife impactor are shown in Fig.4. The specimen heat treated at 850°C exhibits slightly less penetration crack, and the specimen which heat treated at 1030°C, reach the maximum penetration crack as compared with the other targets at the same energy levels. It can be seen that the specimens which heat treated at 850°C and 930°C show better performance against the knife impactor either in group1 or in group 2, especially the specimen heat treated at 850°C.

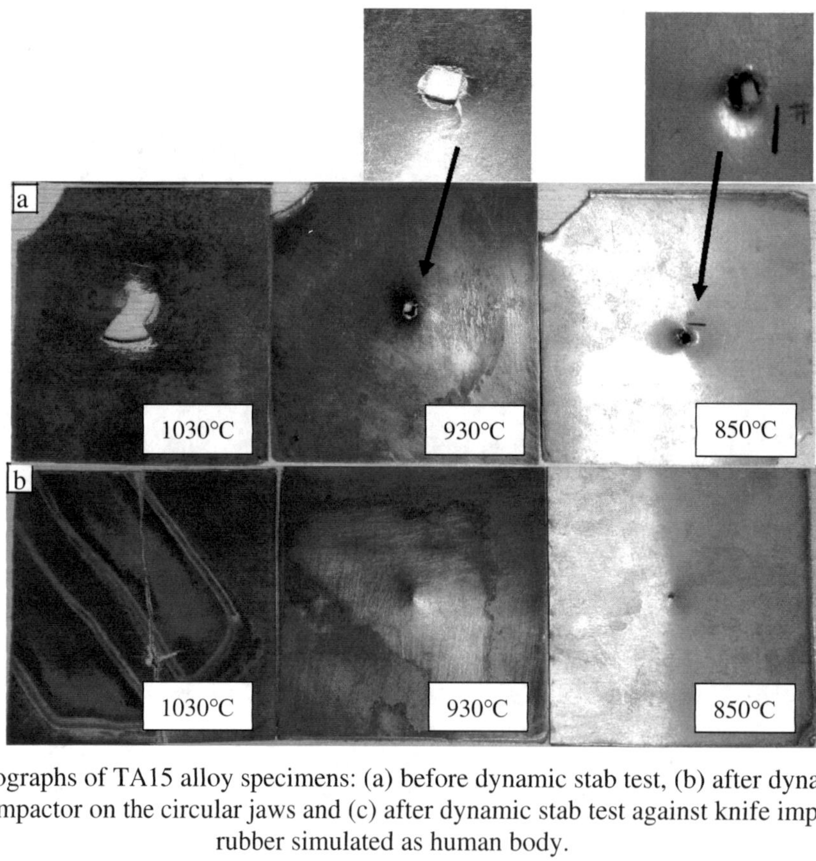

Figure 4. Photographs of TA15 alloy specimens: (a) before dynamic stab test, (b) after dynamic stab test against knife impactor on the circular jaws and (c) after dynamic stab test against knife impactor on the rubber simulated as human body.

It is mean that higher strength and plasticity of TA15 specimens are benefit to the performance of stab resistance. In addition, note that the extent of damage is clearly less in the group 2, although penetration occurs in both targets. In the specimen which heat treated at 930°C,the penetration is a small dot as diameter of about 0.2mm and depth of 2mm, but the penetration is a dot as diameter of about 0.4mm and

0.2mm of depth, and the specimen which heat treated at 1030°C is cracked definitely. It can be deduced that the energy of impact is absorbed by the rubber mainly and the high strength of TA15 specimens is used to resist the strength of stab. So the high strength layer which in the upper or middle location to stab impactor would be more effective for stab resistance than which in bottom in sandwich structure.

3.2. Relationship between property of stab-resistance and microstructure of TA15 alloy

Fig. 5(a) shows the impact force of TA15 alloy specimens against knife impactor during the dynamic stab test on the circular jaws and at the drop height of 0.52 m. It can be seen that all the curves exhibit the peak of impact force, and the curves of 930°C and 850°C exhibit two peaks, but the time ranges are different for specimens with different heat treatment temperature. It is related with the different way of energy absorption during stab resistance. The gradient of impact force and time at the increasing curve shows the process of energy absorption during penetration of stab, and the peak of impact force indicate the amount of energy absorbed by specimens before penetration, lastly the time range of curve reveal the ability of stab resistance for specimen. As can be seen in the Fig.5(a), the curve of 1030°C exhibits the shortest time of about 0.3ms to reach the peak of impact force, and the peak of impact force is 2.55KN when the specimen penetrated. Compared with other specimens, it is also mean the poor performance of stab resistance for specimen heat treated at 1030°C. The curves of 930°C and 850°C exhibit longer time range, show the relatively large amount of energy absorbed and gradually reach the peak of impact force, especially the curve of 850°C. The specimen heat treated at 850°C exhibits the better performance against the knife impactor as proved in Fig.5.

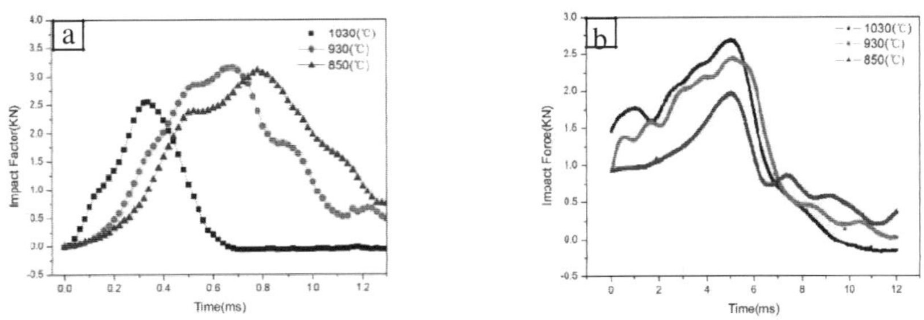

Figure 5. Dynamic stab test of TA15 alloy specimens against knife impactor: (a) on the circular jaws, (b) on the rubber simulated as human body.

In order to analyze the damage mechanisms in the TA15 alloy specimens, the OM and SEM images of TA15 alloy specimens tested against the knife impactor under drop tower conditions are shown in Fig.6.

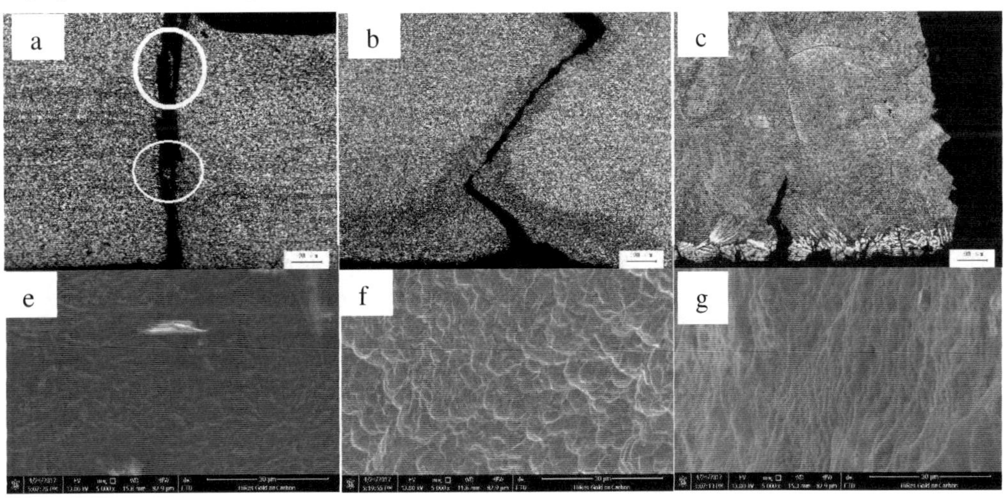

Figure 6. The microstructure nearby the penetration and fracture of TA15 alloy specimens which heat treated at different temperature after dynamic stab test. (a)(e) 850°C, (b)(f) 930°C, (c)(g) 1030°C.

The regions of crack extension in the direction of thickness in specimens are observed by OM, and the fractures are observed by SEM. For the microstructure of TA15 alloy heat treated at 850°C, the extended region of the penetration crack is discontinuous (the region circled by white in Fig.6a)and the fracture is consisted with large amount of fine dimples(in Fig.6e). It is indicated that crack extension occurred in the specimen after the relatively long process of plastic deformation accumulating to some degree. In Fig.6b, the specimen heat treated at 930°C experiences extended crack continuously and large dimples can be seen in fracture as shown in Fig.6f, which reveals that the crack resistance of specimen is related with plastic deformation. Lastly, the SEM photograph of specimen heat treated at 1030°Cshow brittle fracture, and the typical intergranular fracture induced by the stress concentration from the directed contact with the tip of the knife impactor can be observed in Fig.6g.

As mention above, it is obvious that the performance of TA15 alloy during the dynamic stab test against knife threat can be effectively improved by controlling the microstructure, and the specimens with fine grains show stronger capability to absorb the impact force by grain boundary binding force mainly.

4. Conclusions

The dynamic stab resistance properties of TA15 alloy with different microstructure against knife threats were studied in this paper. The relationship between impact force-time curve and mechanical properties show that the properties of higher strength and plasticity are benefit to the performance of stab resistance of TA15 specimens. All the curves of dynamic stab resistance during stab test exhibit the peak of impact force, and the curves of 930°C and 850°C exhibit two peaks and longer time ranges, but the curve of 1030°C exhibits the shortest time to reach the peak of impact force. The research of microstructure and fracture morphology of TA15 alloy with different heat treatment indicate that the specimens heat treated at 850°C exhibit the better performance against the knife impactor because of fine grain and stronger grain boundary binding force to absorb the impact force energy.

5. References

[1] Z H Tan, L Zuo, W H Li, L S Liu and P C Zhai. *Materials and Design*, **94** (2016), p.105
[2] A H Sofiyev and E Osmancelebioglu. *Composites Part B: Engineering*, **120** (2017), p.197
[3] A Serjouei, G Gour, X F Zhang, S Idapalapati and G E B Tan. *International Journal of Impact Engineering*, **105**(2017), p.54-67
[4] Iqbal MA and Gupta NK. *Lat Am J Solids Stru.* **5**(2008),4:259-287.
[5] Haro EE, Szpunar JA and Odeshi AG. *Compos Part Appl Sci Manuf*, **87**(2016), p.54–65.
[6] S Hazama, M Oka and Y Hinoda. Stab Resistance of Aramid Fabric Reinforced with Waterborne Polyurethane Resin. *Nippon Rinsho Japanese Journal of Clinical Medicine*, Vol.**50**(2013), p.252-257
[7] H Kim and I Nam. *Journal of Applied Polymer Science*, **123**(2012), p.2733
[8] Jianjun Zhua, Pei-Zhou Li, Wenhan Guo, Yanli Zhao and Ruqiang Zou. Titanium-based metal–organic frameworks for photocatalytic applications. *Coordination Chemistry Reviews*, Vol.**359**(2018), p.80
[9] Ivan Smirnov, Alexander Polyakov and Yuri Sudenkov, Strength and fracture of ultrafine-grained titanium Grade 4. *Procedia Structural Integrity*, Vol. **6**(2017), p.196
[10] Zhichao Sun, Huili Wu, Junzhi Sun and Jing Cao. *International Journal of Hydrogen Energy*, Vol. **42**(2017), p.20849
[11] Pengfei Gao, Mei Zhan, Xiaoguang Fan, Zhenni Lei and YangCai. *Materials Science and Engineering* A, Vol.**689**(2017), p. 243
[12] Stab resistance of personal body armor. national institute of justice standard 0115.00; 200

IOP Conf. Series: Materials Science and Engineering **838** (2020) 012007 doi:10.1088/1757-899X/838/1/012007

Influence of solution temperature on microstructure and mechanical properties of wrought Al-Si-Mg alloy

F Liu, F Gao, F X Yu, D Z Zhao

School of Materials Science and Engineering, Northeastern University, Shenyang 110819, People's Republic of China

Email: liufang@smm.neu.edu.cn

Abstract. The microstructure and mechanical properties of wrought Al-12.7Si-0.7Mg alloy were investigated aiming at understanding the influence of the solution treatmen. The result show that the Si particles were gradually rounded and the grain size increase with increasing the solution temperature from 500°C to 540°C. The alloy in T4 condition changed due to the grain szie, Si particles and the super saturation level of Mg ans Si solute atoms. With higher solution temperature, the volume fraction of the pricipitate becomes more which result in good manchannical properties.

Keywords: Wrought Al-Si-Mg alloy, Solution treatment, Microstructure, Mechanical property

1. Introduction

The main alloying elements in 6xxx Al-Mg-Si aluminum alloys and wrought Al-Si-Mg alloys are Si and Mg, which form Mg_2Si phase in the equilibrium state and a series of transient phases as precipitation during aging treatment following solution treatment [1-4]. Al-Si alloys with addition of Mg make a big group of cast aluminum alloys for structural purpose.

The fact that the mechanical properties of the Mg bearing Al-Si alloys can be made comparable or even superior to that of the 6000 series alloys through traditional metallurgical process is attractive from the viewpoint of industrial application of the Al-Si alloys [5]. The strengthening mechanisms are from precipitation, grain boundary and particulate strengthening. The main feature of wrought Al-Si-Mg alloy is that it contains uniformly distributed silicon particles in the Al matrix of fine equiaxed grains.

During heat treatment process, the normal grain growth of Al matrix and the coarsening of the Si particles at high temperature might be restrained by mutual reactions. In this paper,The microstructure and mechanical properties of the alloy were investigated aiming at elucidating the impact of the solution treatment.

2. Material and methods

DC cast Al-12.7Si-0.7Mg billet without chemical modification in Ref. [5] was used in this study. After pre-heat-treatment at 480°C for 3 hours, the billets were extruded into a profile of 4 mm in thickness at the same temperature.

The extruded samples were solution-treated at 500°C, 520°C and 540°C for 30 min, and subsequently quenched into water. The quenched samples were artificial aged at 180°C for 3 h in a Muffle furnace. The tensile specimens and wear test specimens were prepared from extruded profile along the extruding direction. Specimens were tensile-tested at a cross-head speed of 1 mm/min on a SANS testing machine.

IOP Conf. Series: Materials Science and Engineering **838** (2020) 012007 doi:10.1088/1757-899X/838/1/012007

An optical microscope (Leica DMI 5000M) and a FEI Tecnai F20 transmission electron microscope (TEM) were used to observe the microstructures of samples. Image-Pro Plus image analysis software was used for investigation of the Si particle size distribution in the alloy and the grain size of the Al matrix size.

3. Results and discussion

Fig. 1 show the microstructure of longitudinal (along the extrusion direction) of Al-12.7Si-0.7Mg alloy in T4 (solution treatment+natural aging) and T6(solution treatment+artificial aging), respectively. It can be seen that a number of Si particles distributed in Al matrix and the size of Si particle increase by varying the solution temperature. Aging at 180°C, the size of Si particle is not changed by the aging treament. The Si particles and the grain sizes were obtained by measuring as shown in Table 1. It is noticed that the Si particles were gradually rounded and their aspect ratio were decreased in average. It is worthy to notice that the Al matrix grains increased to 5.5μm increasing solution treatment temperature to 540°C which result in the derease of YS and UTS.

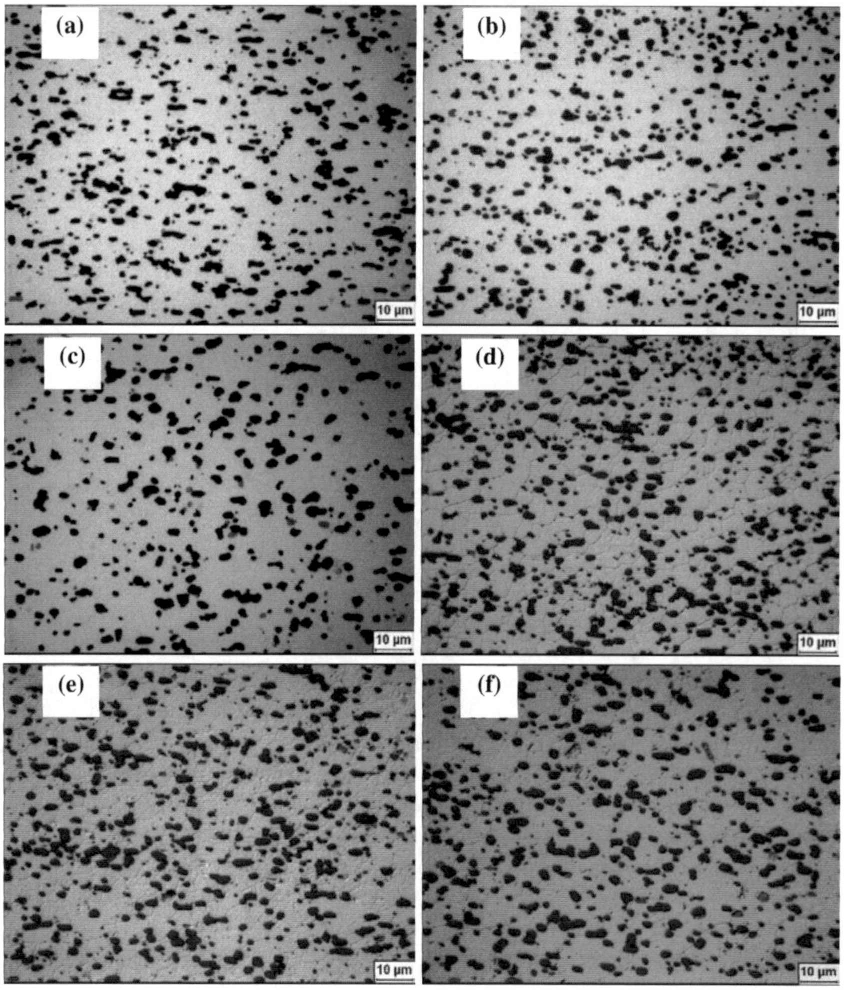

Figure 1. Optical micrographs of extruded alloy in different conditions 500°C×30min, (b)520°C×30min, (c)540°C×30min, (d) 500°C×30min+180°C×3h, (e) 520°C×30min+180°C×3h,(f) 540°C×30min+180°C×3h.

Table 1. The size of the Si particle and grain of the alloy in different condition.

Alloy statements	Si particles size（μm）	Aspect ratio	Grain size（μm)
500°C×30min+180°C×3h	2.8	1.91	5.1
520°C×30min+180°C×3h	3.1	1.70	5.2
540°C×30min+180°C×3h	3.0	1.57	5.5

Table2 lists the mechanical properties of the Al-12.7Si-0.7Mg alloy treated in different conditions. As shown in Table 1, yield strength (YS), ultimate tensile strength (UTS) and Elongation of the alloy in T4 and .T6 tempers are much higher than that in T1 (as extruded). The optical microstructures shown in Fig.1 indicate that there are merely differences between that in T4 condition and that in T6 conditions. On the other hand, it is thus clear that precipitation hardening makes contribution to the increased YS and UTS in T6 treated alloy[6-8].

After solution treatment, the solubility of Mg_2Si phase increases with increasing of solution temperature from 500°C to 540°C. The Mg_2Si phase is in the form of solute atoms into the matrix which act as solid-solution strengthening. The YS and UTS varies directly with introducing Mg and Si solute atoms. With more Mg and Si solute atoms into the matrix, and the strength of the alloy is higher[9]. Compared to 520°C, YS and UTS of the alloy decrease when heat treated at 540°C because the equiaxed Al grain size increase to from 5.1 to 5.5μm.

Fig. 2 shows the bright-field TEM micrograph (B=$[001]_{Al}$) of the alloy in T6 condition. The TEM images of the T6 condition alloy are characterized by very fine dot and needle precipitates. The needle-shaped precipitates are oriented along $[010]_{Al}$ and $[100]_{Al}$ directions. It can be seen that volume fraction of the precipitates is higher in Fig. 2(c) due to the more solute atoms in the matrix to form the precipitates.

Table 2. Mechanical properties of the Al-12.7Si-0.7Mg alloy in T1, T4 and T6.

Alloy statement	YS (MPa)	UTS(MPa)	Elongation(%)
T1(as extruded)	108	190	15.0
500°C×30min	136	263	16.2
520°C×30min	162	292	13.7
540°C×30min	148	285	15.9
500°C×30min+180°C×3h	250	313	7.4
520°C×30min+180°C×3h	270	333	7.3
540°C×30min+180°C×3h	293	358	6.7

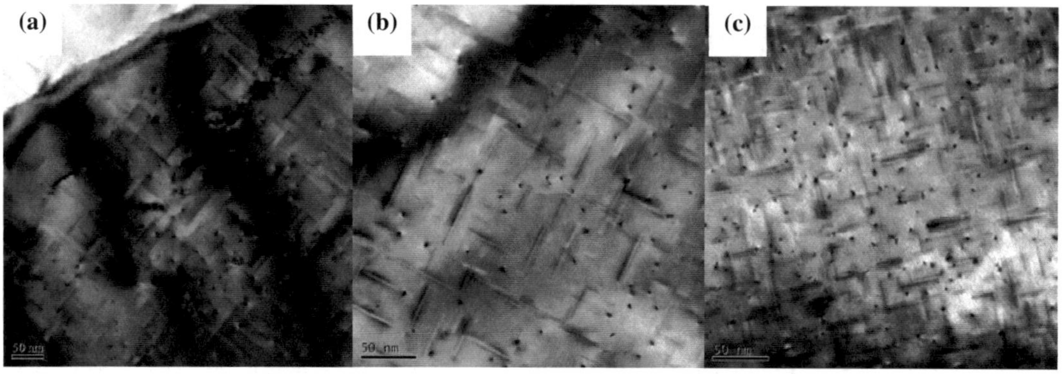

Figure 2. TEM images of the alloy at different treatment conditions: (a) 500°C×30min+180°C×3h, (b) 520°C×30min+180°C×3h, (c) 540°C×30min+180°C×3h.

4. Conclusion

(1) The Si particles were gradually rounded and their aspect ratio were decreased in average and the grain size increase with increasing the solution temperature from 500 °C to 540°C
(2) Wrought Al-12.7Si-0.7Mg alloy in T4 condition have good mechanical properties due to solid-solution strengthening.
(3) Elevating the solution treatment temperature can effectively enhance the volume fraction precipitation phase.

5. References

[1] Troeger L P and Starke Jr E A *Mater. Sci. Eng.* A **277** (2000), p. 102
[2] Andersen S J, Zandbergen H W, Jansen J, Traeholt C, Tundal U and Reiso O *Acta Mater.* **46** (1998), 3283.
[3] Marioara C D, Andersen S J, Jansen J and Zandbergen H W *Acta Mater.* **49** (2001), 321.
[4] Fatahalla N, Hafiz M and Abdulkhalek M *J. Mater. Sci.* **34** (1999), 3555.
[5] Fang L, Fuxiao Y, Dazhi Z and Liang Z *Mater. Sci. Eng.* A **528** (2011), 3786.
[6] Miao W F, Laughlin D E *Scripta Mater.* **40** (1999), 873.
[7] M A Munoz Morris, I Gutierrez-Urrutia and D G Morris *Mater. Sci. Eng.* A **493**(2008), 141.
[8] Bao G, Hutchinson J W and McMeeking R M *Acta Metall. Mater.* **39**(1991), 1871.
[9] Congchang X, Hong H, Wanyu Y and Luoxing L *Mater. Sci. Eng.* A **744**(2019), 28.

Acknowledgments

This work was supported by National Key R&D Program of China (2017YFB0306402), the National Natural Science Foundation of China under grant No. 51401047 and by the Fundamental Research Funds for the Central Universities No. N130302002.

Finite element model and experimental verification of Glare laminate impact

X CUI[1, a], Z Y Zhang[2,b] , S Wang[2,c]

[1]College of Civil Aviation, Shenyang Aerospace University, Shenyang, China
[2]College of Aerospace Engineering; Shenyang Aerospace University, Shenyang, China

Email: [a]772528788@qq.com, [b]285226827@qq.com, [c]wangshuo2552@sina.com

Abstract. In this paper, a finite element simulation model is established for the impact of Glare laminates, and the reliability of the model is verified by experiments. Glare applies the continuum damage mechanics (CDM) model and inputs the user material subroutine VUMAT to determine the damage of the material based on the theory of CDM under plane stress state. By comparing the ultrasonic scanning experiment after impact damage with the CDM finite element simulation cloud pattern used in this paper, the constitutive relation of the fibre continuous damage model used in this paper can better simulate the damage initiation, evolution and damage of internal materials.

Keywords: Glare; CDM; VUMAT in Abaqus; Low speed impact

1. Introduction

In order to improve the damage tolerance of aviation structures and reduce the quality of structures, the aviation industry has been devoting itself to the development of new materials. Glare is a new generation of fibre reinforced metal laminates developed to solve the problem of fibre breakage in ARALL under fatigue loading [1]. It is made of 0.3-0.5mm thick aluminum alloy sheet and 0.25-0.5mm thick unidirectional or bi-directional glass fibre reinforced composite material laying and pressing in a specific order. The prepreg layer is bonded with epoxy resin and glass fibre. The number of layers can be selected according to the design requirements [2].

As a new practical composite material with super-hybrid structure, the failure mode of Glare laminates under impact is relatively complex [3], which not only includes metal layer cracking and fibre layer cracking[4], but also has an important failure mode, namely interlaminar debonding. Therefore, the simulation of the impact of Glare is complicated. In the theoretical study and numerical simulation of interfacial debonding, many new theoretical models and methods have been proposed by scholars at home and abroad. For example, Jiang et al [5] proposed a concise constitutive model for interface debonding, which describes the interface cracking process in mixed mode. The simulation results in finite element software LS-DYNA are in good agreement with the experimental results. Turon et al [6] proposed a thermodynamic damage constitutive model of layered interface under thermo-mechanical coupling. The model is based on damage mechanics theory, fully considering the thermodynamic effect of interface, and avoids the interpenetration of adjacent layers after debonding. Compared with the experimental results, the numerical prediction has better results. Li Biao et al [7] proposed a new interface element for simulating interface debonding, which combines rigid element and zero-thickness cohesion element [8]. On the premise of ensuring the accuracy of calculation, it can reduce the number of elements in the interface crack tip area, reduce the calculation scale and improve the calculation efficiency. Using this element, the interface cracking of FMLs can be well simulated in the simulation of mixed bending test and double cantilever beam tension test [9].

Based on previous studies, this paper conducts finite element modeling and simulation of low-speed impact of GALARE laminates by abaqus under the theory of continuous damage medium.

Finally, the finite element model is verified by experiments.

2. Finite element simulate method

In this study, for the finite element modeling of Glare laminates subjected to impact, we divide the Glare laminates micro-structure into three parts: fibre reinforced composite layer, metal layer and bonding layer[10].

For fibre composite layers in Glare. By comparing various failure , such as the classical HASHIN failure criterion.The theoretical model of Continuous damage MODLE is selected[11].

CDM mechanics includes three parts: damage characterization, damage determination and damage evolution [12]. Corresponding to different damage modes of composite materials, damage characterization establishes the constitutive relationship between damaged materials and intact materials by introducing damage state variables [13]. The finite element type is C3D8R Three-dimensional solid element. For metal layers in Glare, The conventional Jonson cook constitutive model is chosen for the isotropic metal layer. Find and use the properties of 5052 aluminium alloy panel. Cohesive zone theory is adopted in the bonding layer, and cohesive element built in ABAQUS is used to realize the bonding between layers [14].

The simulation process of the failure problem is realized by adding damage subroutine into ABAQUS finite element software. The model (three-dimensional solid model) uses Shokrieh's improved Hashin failure criterion to judge the failure mode of the model, and degrades the material properties [15]. Four failure modes are considered in the model: matrix tensile or compressive failure, fibre tensile or compressive failure, fibre matrix shear failure, normal tensile or compressive failure [16].

According to the above failure criteria, the progressive damage judgment process as shown below can be summarized:

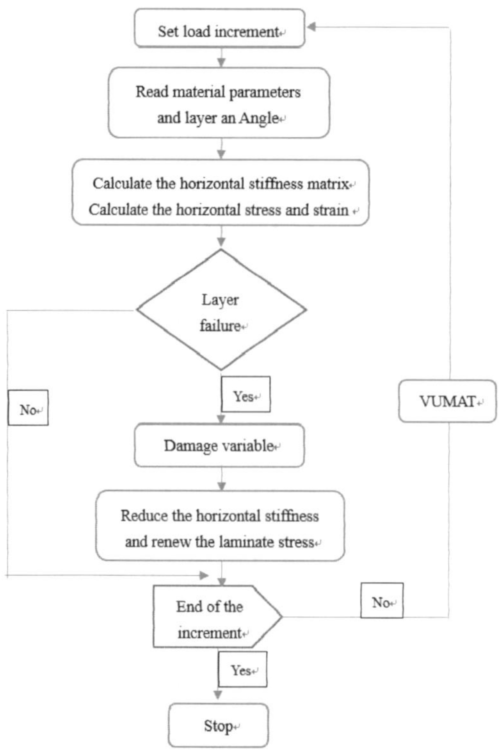

Figure 1. Progressive damage assessment procedure.

3. Manufacturing and impact testing

3.1. Part of manufacturing

Glare5 types laminates were used in this study, laminates consist of four 0.3 mm thick 2024-T3 aluminium layers, bonded together with glass fibre prepregs S2-glass/X1101. In Glare5, each glass prepreg laminate between the aluminium plates is made of uni-directional (UD) plies with a layup of [0/90/90/0].The size of prepreg (Fig 2a) and aluminium alloy sheet cut to 110 cm × 110 cm.

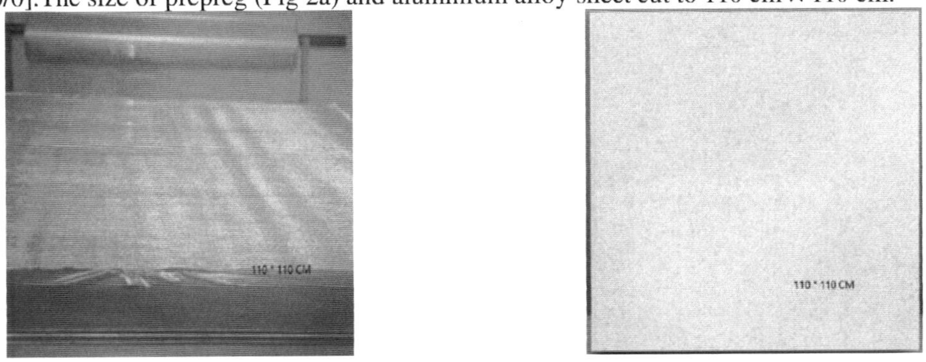

a. Prepreg (S2-glass/X1101) cut into suitable size

b. The surfaces pre-treated aluminium plates cut into suitable size

Figure 2. Basic materials and pretreatment.

Prior to bonding, the aluminium surfaces were pre-treated with chromic acid anodizing and primed with X1101 (Engineered Materials, Key Laboratory of Advanced Polymer Matrix Composites in Liaoning Province, China). The surfaces pre-treated aluminium plates are shown in the figure 2b.

Glare plates are manufactured using Hot-pressed Autoclave in the Composite Materials Laboratory of Shenyang Aerospace University, China, Showing in fig 3. Hot-pressed Autoclave links to the digital operating table, which can be heated according to the set temperature curve, and can also be pressurized according to the set pressure curve.

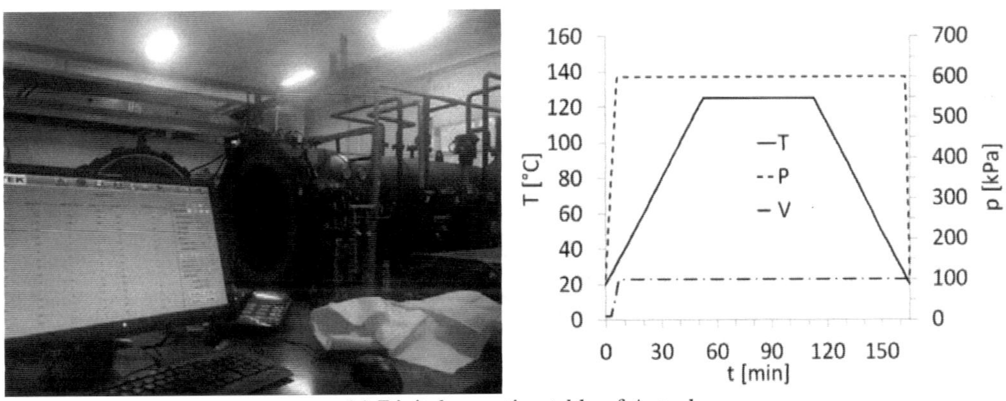

(a) Digital operating table of Autoclave

(b) Standard manufacturing conditions for Glare panels during the cure cycle: Temperature (T), pressure in the vacuum bag (V) and pressure in the autoclave (P)

Figure 3. Hot-pressed Autoclave Manufacturing.

3.2. Part of impact testing

The standard drop hammer impact test bench is used in the experiment, including fuselage, frame, upper platform and lower platform. The chassis is installed on the upper platform. The chassis consists of a box-shaped frame, a driving stepper motor, two lead screws and two polished rods. The fuselage consists of two parallel guide pipes fixedly connected to the bottom of the mounting pedestal, a driving device for adjusting the impact height, a hammer body and an anti-secondary impact device. Fig 4 is standard drop hammer impact test bench.

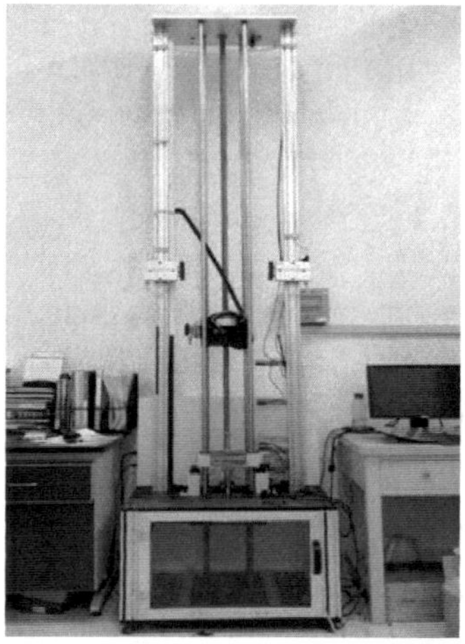

Figure 4. Standard drop hammer impact test bench.

After the finite element simulation, corresponding experiments are set up to verify the simulation results. In order to study the damage of glitches caused by low-speed impact, we set different impact energies and respectively tested the impact damage of blazes at different energies. Different impact energy can be achieved by changing the height of the test bench punch. The height of the punch is 0.1 to 2.3 m, the mass of the punch is 2.6 kg, and the tip radius of the punch is 8 mm. The impact energy of the punch is 10.22j, 12.38j and 14.46j. The impact velocity is calculated from the kinetic energy of the punch [17].

4. Results and discussion

Metal layered observation results, such as Fig 5 shows that upper Aluminum layer (a) and lower Aluminum ply layer (b). It can be observed that the plastic deformation of the uppermost metal laminate is evenly distributed throughout the laminate, while the plastic deformation of the lowermost metal laminate is concentrated at the impact point. Similarly, we can observe that in the composite layer, the degree of matrix crack and interface damage also increases gradually from top to bottom.

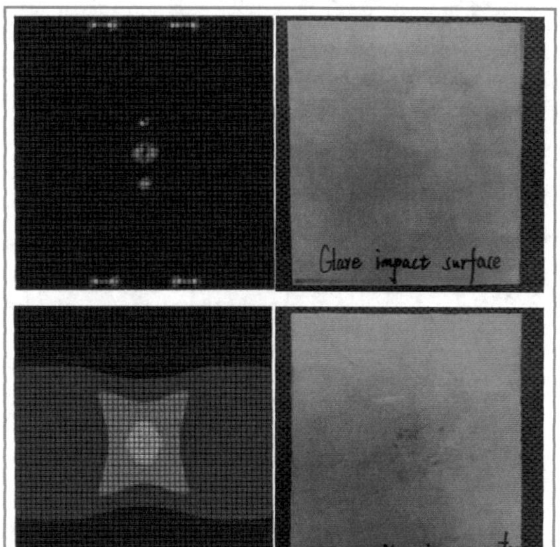

(a) upper Al layer (b) lower Al ply layer

Figure 5. Damage model: plastic dissipation in Aluminum.

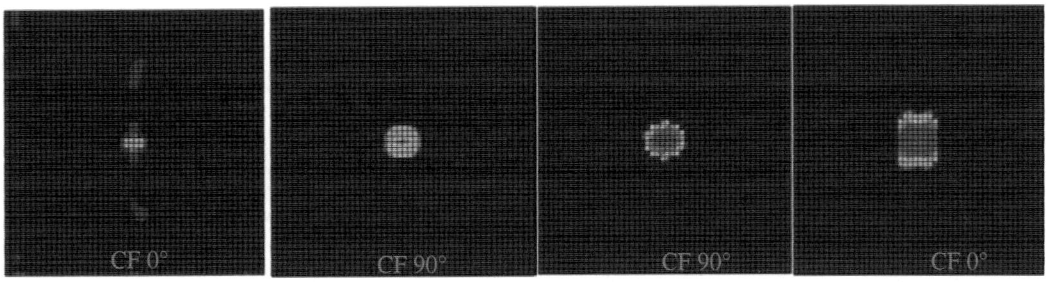

Figure 6. Damage model: matrix cracks in composite (CF) plies and debonding in interface.

As shown in the following Figure 6, the bottom illustration shows four layers of composite laminates from top to bottom. It shows areas of plastic deformation in Aluminum l layers, debonding in cohesive interface elements and matrix failure in composite plies. These areas are mainly concentrated around the impact. Some plastic dissipation owing to clamping can be observed in the upper Aluminum layer. More damage is observed in the lower plies owing to bending. The damage area in the lower Aluminum ply and the lower interface layer is significantly larger than in the corresponding upper layer.

5. Conclusion

In this paper, for the Glare laminate, we have done the finite element simulation and experimental verification of the low-speed drop hammer impact test, and finally reached the relevant conclusions. The continuous damage model (CDM) constitutive relation of the fibre used in this paper simulates the damage initiation, evolution and damage of the internal material better, and provides a dynamic response for the predicted Glare plate under impact load. A good theoretical basis. From the comparison of the ultrasonic scanning experiment and the finite element simulation cloud image after the impact damage, the tensile stress of the 90° layer of the Glare plate is greater than the tensile stress of the 0° layer, and the compressive stress of the 0° layer is greater than 90°. Compressive stress; and the tensile stress on the back side of the Glare board is particularly prominent.

6. References

[1] Vermeeren C A J R. An Historic Overview of the Development of Fibre Metal Laminates[J]. *Applied Composite Materials*. 2003, 10(4-5) 189-205.
[2] Schijve J Development of fibre-mental laminates, arall and Glare, new fatigue resistant materials, LR-715[R]. Delft: Faculty of Aerospace Engineering, Delft University of Technology, 1993
[3] Sinke J. Manufacturing of Glare Parts and Structures[J]. *Applied Composite Materials*. 2003, 10(4-5) 293-305.
[4] Donadon M V, Iannucci L, Falzon B G, et al. A progressive failure model for composite laminates subjected to low velocity impact damage[J]. *Computers & Structures*. 2008, 86(11) 1232-1252.
[5] Jiang W G, Hallett S R, Green B G, et al. A concise interface constitutive law for analysis of delamination and splitting in composite materials and its application to scaled notched tensile specimens[J]. *International Journal for Numerical Methods in Engineering*, 2010, 69(9) 1982-1995.
[6] Turon A, Camanho P P, Costa J, et al. A damage model for the simulation of delamination in advanced composites under variable-mode loading [J]. *Mechanics of Materials*. 2006, 38(11) 1072-1089.
[7] Biao Li, et al. Characterization of a combined interfacial element for laminated composites [J]. *Acta material*. 2013, 30(5) 159-165.
[8] Carrillo J G and Cantwell W J. Scaling Effects in the Low Velocity Impact Response of Fibre-Metal Laminates[J]. *Journal of Reinforced Plastics & Composites*. 2008, 27(9) 893-907.
[9] Linde P, Pleitner J, Boer H D, et al. Modelling and Simulation of Fibre Metal Laminates[J], 2004.

[10] Maimí P, Camanho P P, Mayugo J A, et al. A continuum damage model for composite laminates: Part II – Computational implementation and validation[J]. *Mechanics of Materials*. 2007, 39(10) 897-908.

[11] Ladeveze P and Ledantec E. Damage modelling of the elementary ply for laminated composites[J]. *Composites Science & Technology*. 1992, 43(3) 257-267.

[12] Maa R H and Cheng J H. A CDM-based failure model for predicting strength of notched composite laminates[J]. *Composites Part B Engineering*, 2002, 33(6) 479-489.

[13] Turon A, Camanho P P, Costa J, et al. A damage model for the simulation of delamination in advanced composites under variable-mode loading [J]. *Mechanics of Materials*, 2006, 38 (11) 1072-1089.

[14] Hwang W C, Sun C T. Failure analysis of laminated composites by using iterative threedimensional finite element method[J]. *Computers & Structures*, 1989, 33(1) 41-47.

[15] Tay T E, Liu G, Tan V B C, et al. Progressive Failure Analysis of Composites[J]. *Journal of Composite Materials*, 2008, 42(18) 1921-1966.

[16] Zhang Y, Zhu P, Lai X. Finite element analysis of low-velocity impact damage in composite laminated plates[J]. *Journal of Shanghai Jiaotong University*, 2006, 27(6) 513-519.

[17] Cui X, LI B, Shuo W et al. Simulation on low velocity impact to Glare laminate based on ABAQUS[J]. *Journal of Aeronautical Materials*, 2019, 39(02) 72-78.

Acknowledgments

Foundation item: Overseas Training Foundation For University of LiaoNin g Province of China （2018 LNGXGJWPYYB008）Liaoning Provincial Natural Science Foundation of China (No.2019-ZD-0246)

IOP Conf. Series: Materials Science and Engineering **838** (2020) 012009 doi:10.1088/1757-899X/838/1/012009

Numerical study on effect of micro-amplitude vibration on the grains morphology of alloys

Lei Rao*, Kai You, Xue Bai and Tianyu Miao
College of Mechanics and materials, Hohai University, Nanjing 211100, China

*corresponding author: raol@hhu.edu.cn

Abstract. Combining with CAFE model, this paper introduced a modified heat transfer model which interfere the effect of mechanical vibration based on turbulent heat transfer theory, on this basis the grains morphology of alloys under micro-amplitude vibration can be analyzed. Based on this model, the results of ZL201 alloys showed that vibration can promote columnar-to-equiaxed transition (CET) and grain refinement, and improve homogeneity of grains distribution. The optimum vibration frequency in current study is 40Hz. The method used in this paper provides a reference for study on the theory of vibration casting and is beneficial for parameters optimization of mechanical vibration in foundry field.

1. Introduction

Grain refinement for metallic materials is a very efficient way to obtain outstanding performance including high strength, excellent bend-ability and high ductility. The application of sonic, ultrasonic and mechanical vibration can promote grain refinement and change the distribution form of second phase [1, 2]. Among above methods, application of mechanical vibrations attracts much attention because of a simple system [3].

The formation of microstructure of castings results from the nucleation and growth of grains. Nucleation and growth of grains depend on the state of under-cooling liquid melt, which is determined by the solidification rate and temperature gradient of dendrite tip. Cellular automaton-finite element (CAFE) model has been successfully used to simulate the solidification microstructure under various casting conditions in recent years [4, 8]. However, the method has rarely been used in vibration casting.

Combining with CAFE model, a modified heat transfer model was established in this paper which considered the effect of mechanical vibration on casting. On this basis grains morphology of alloys can be analyzed, which provides a reference for parameters optimization of vibration casting.

2. Mathematical model

2.1. Heat transfer equations

Under micro-vibration amplitude condition, the Reynolds number in the molten pool is very small, so heat transfer in the melt is mainly in the way of heat conduction, which can be expressed as:

$$\rho c_p \frac{\partial T}{\partial t} = \frac{\partial}{\partial x}\left(\lambda \frac{\partial T}{\partial x}\right) + \frac{\partial}{\partial y}\left(\lambda \frac{\partial T}{\partial y}\right) + \frac{\partial}{\partial z}\left(\lambda \frac{\partial T}{\partial z}\right) + Q_V . \tag{1}$$

Where ρ is the density; c_p is the specific heat; T is the temperature; Q_V is the energy input by vibration; λ is effective thermal conductivity, which can be expressed by:

$$\lambda = \lambda_m + \lambda_t \,; \quad \lambda_t = \frac{1}{P_{rt}} \rho C_p C_\mu \frac{\kappa^2}{\varepsilon} . \tag{2}$$

Where λ_m is static thermal conductivity, λ_t is turbulent thermal conductivity [5]. ε is turbulent flow dissipation rate, κ is pulsating kinetic energy. C_μ is experimental constant and takes 0.09. Considering the disorder of liquid motion under micro-amplitude vibration condition, three-dimensional vibration energy Q_{Vi} is exerted on casting mould to weaken the effect of vibration direction. The same frequency f and amplitude A are exerted at each orthogonal coordinate direction, and the difference of phase angle between each two coordinate axis is 120 degrees.
The total vibration energy Q_V can be expressed as:

$$Q_V = 3Q_{Vi}; \ Q_{Vi} = 8\pi^2 \rho A^2 f^3, \ i = x, y \text{ and } z . \tag{3}$$

Liquid flow in the bulk melt is a blocked state when the solid fraction g_S is higher than 0.4 [6-7], in this case $Q_V = 0$ and $\lambda_t = 0$.

2.2. CAFE model

2.2.1. Nucleation equations

It is assumed that nucleation become active when the under-cooling increases to an active value. In present work, continuous nucleation model proposed by Rappaz and Gandin [4] is applied. The relationship between nucleation density and under-cooling is given by:

$$\frac{dn}{d(\Delta T)} = \frac{n_{max}}{\sqrt{2\pi}\Delta T_\sigma} exp\left[-\frac{(\Delta T - \Delta T_{max})^2}{2\Delta T^2}\right] . \tag{4}$$

The grain density can be calculated by equation:

$$n(\Delta T) = \int_0^{\Delta T} \frac{dn}{d(\Delta T)} d(\Delta T) . \tag{5}$$

Where n is the grain density, ΔT is the under-cooling, n_{max} is the maximum nucleation density, ΔT_σ is the deviation standard of the nucleation under-cooling, and ΔT_{max} is the maximum nucleation under-cooling.

2.2.2. Dendrite growth equations

The total under-cooling of the dendrite tip ΔT is the sum of four contributions [8-9]:

$$\Delta T = \Delta T_c + \Delta T_t + \Delta T_k + \Delta T_r . \tag{6}$$

Where ΔT_c, ΔT_t, ΔT_k and ΔT_r are the under-cooling value of solute diffusion, thermal diffusion, attachment kinetics, and solid/liquid interface curvature, respectively. For most metallic alloys, the last three under-cooling are very small and can be ignored, thus a simplified calculation function is introduced based on the KGT model. The dendrite tip growth kinetics in the casting will be calculated by:

$$v = a_2 \Delta T^2 + a_3 \Delta T^3 . \tag{7}$$

Where a_2 and a_3 are coefficients of the growth velocity of the dendrite tips, which were 3.12e-007 and 7.56e-008 in this paper, respectively.

2.3. Model and calculation of parameters

Finite element models of cylindrical ingot (with radius of 25mm and height of 140mm) and casting mould (with thickness of 10mm) were established. Standard ZL201 aluminum alloy and H13 were respectively assigned to the ingot and casting mould. The initial temperature of mould and the pouring temperature were 80 °C and 710 °C, respectively. And the heat transfer coefficient of ingot/mould was 2000 W/m^2-K.

3. Results and discussion

3.1. Temperature field and grain morphology

The simulated results of temperature field of castings without and with vibration (10 Hz, 0.2mm) are coupled in figue.1. Temperature fields corresponding to solid fraction of 30% and 60% were selected for both castings. It can be seen that positive temperature gradient was established towards the inner melt during the solidification. Lower temperature gradient was formed in the casting with vibration than that of casting without vibration, which suggests the enhancement of heat transfer in liquid melt.

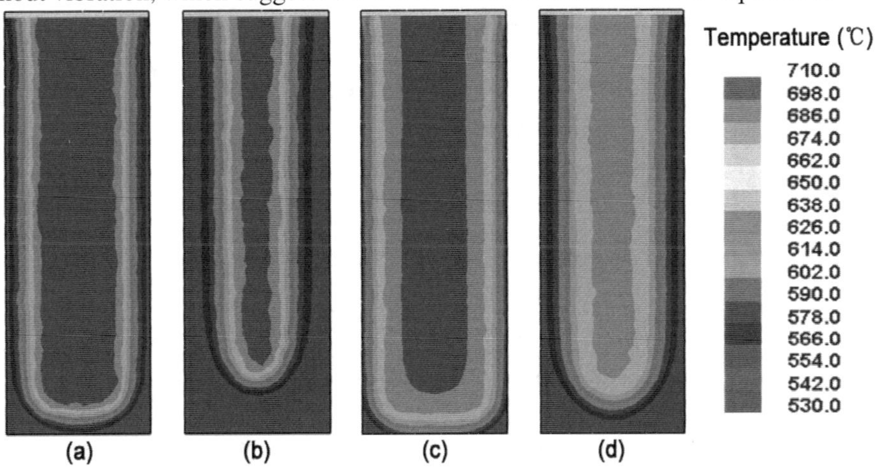

Figure 1. Temperature field of casting without ((a) ,(b)) and with vibration ((c) ,(d)) .(a) $g_S = 30\%$; (b) $g_S = 60\%$; .(c) $g_S = 30\%$; (d) $g_S = 60\%$;

The solidified structure of casting depends critically on the temperature distribution in the molten pool [10, 11]. Figue.2 (left part) shows the grains morphology of the ZL201 alloy castings with and without vibration. As for casting without vibration, wide grains zone near the side wall showed obvious tendency of columnar shape which faced the axis of the cylinder casting. While for the casting with vibration, grains morphology became more disordered and the columnar grain zone was compressed to a very narrow width near the side wall.

The grains morphology mainly depends on the *G/R* ratio [12-13], where *R* is solidification rate and *G* is temperature gradient. High *G/R* ratio promotes formation of columnar grains, while Low *G/R* ratio favors formation of equiaxed grains. Figue.2 (right part) shows the *G/R* ratio of castings with and without vibration. The width of high *G/R* zone in casting with vibration was narrower than that of casting without vibration, thus columnar grain zone of casting with vibration was thinner. The results indicate that vibration can effectively restrain the growth of columnar grains zone.

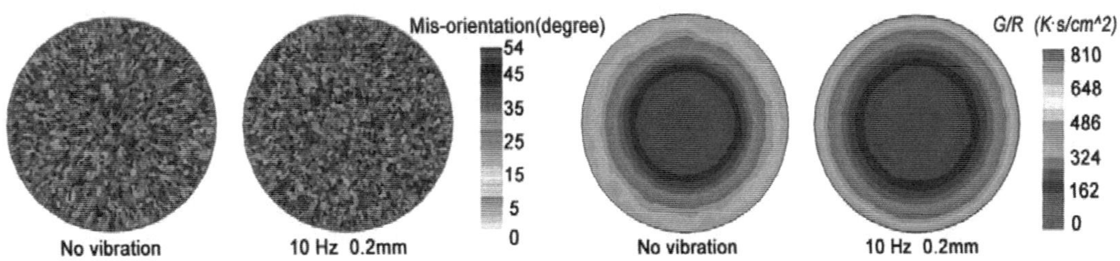

Figure 2. Grain morphology (left part) and *G/R* ratio (right part) of casting.

The homogeneity and compactness degree can be respectively evaluated by variance of grain size and the average radius of grains. Lower variance of grain size represents good homogeneity and smaller radius of grains means higher compactness degree. The distribution of grains radius of the two castings was shown in figure 3. The average grain size of casting with vibration is slightly smaller than that of casting without vibration, showing an improved compactness degree. The grains size mainly depends on the N/K ratio [13], where N is the nucleation density, and K is grain growth rate. With solidification processing, the under-cooling zone are enlarged by vibration, which resulted in a rise of nucleation quantity in under-cooling zone, thus decreasing the grains size of casting. In addition, the variance of grain size in casting with vibration was only 2/3 of that of casting without vibration, which exhibited a higher homogeneity. The improved grain homogeneity may be attributed to lower temperature gradient of under-cooling zone under vibration condition.

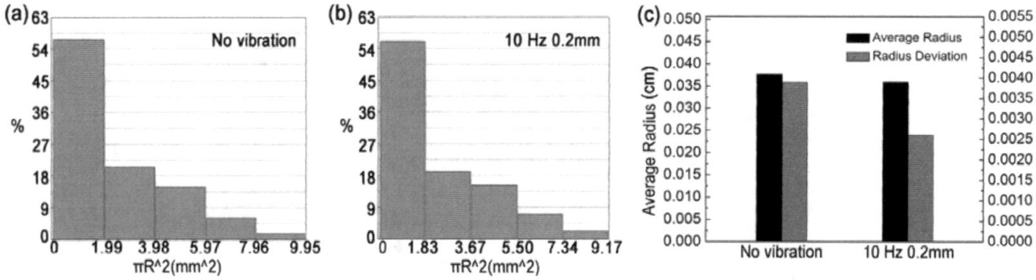

Figure 3. Grain size distribution and statistics of casting with and without vibration.

3.2. Effect of vibration frequency

Figure 4 displays the grains size and variance of castings with different vibration frequency. The average grains radius decreased from 0.036 to 0.0345 cm when vibration frequency increased from 10 to 40 Hz, which owned to the increasing vibration effect on heat transfer. But when the vibration frequency rose to 50 Hz, the average grain size increased to 0.0356 cm, which may caused by the excessive vibration energy. The trend of grain size change is similar to experiment results in reference [14]. Besides, the uniformity of grains became better gradually with increasing vibration frequency, which was attributed to improved uniformity of temperature field in under-cooling zone.

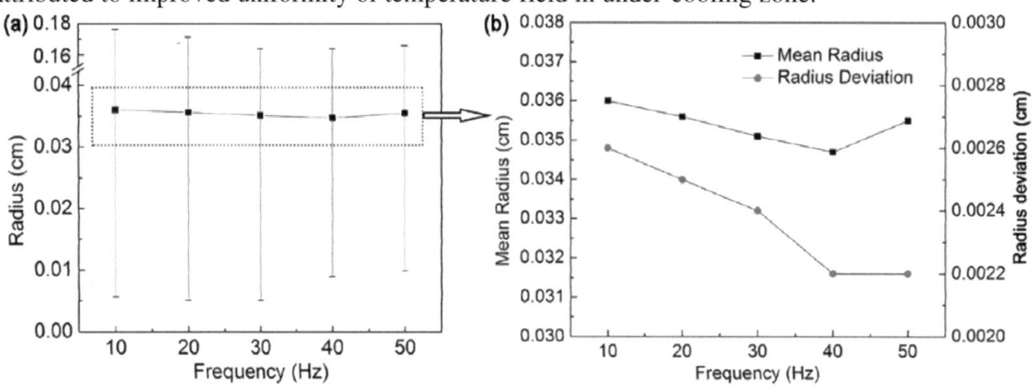

Figure 4. Simulated grain size and deviation of casting with different vibration frequency.

It can be seen that appropriate vibration frequency is beneficial for both grain refinement and uniformity of microstructure, which in turn affect the mechanical properties such as strength and ductility. Highly similar to the conclusion in reference [3], all castings with vibration showed better homogeneity and smaller grain size than casting without vibration, which suggested enhanced mechanical properties.

4. Conclusion

This paper established a modified heat transfer model which considered the micro-amplitude vibration effect on solidification of alloys based on turbulent heat transfer theory. On this basis the grain morphology of alloys can be analyzed combining CAFE model. The simulated results of ZL201 alloys showed that vibration promotes CET and grain refinement, the best vibration frequency is 40Hz. Besides, with increasing vibration frequency, the homogeneity of grain distribution was improved gradually. The improvement of grain morphology was determined by the combined action of temperature gradient, solidification rate and nucleation density, which was influenced by redistribution of temperature field under mechanical vibration condition. The model is significant for research on theory of vibration casting and optimization of vibration parameters in casting.

5. References

[1] Rasgado M T A and Davey K. The effect of vibration on surface finish for semisolid and cast components[J]. *Journal of Materials Processing Tech* 2002, 125–126(13) 543-548.
[2] Kocatepe K. Effect of low frequency vibration on porosity of LM25 and LM6 alloys[J]. *Materials&Design* 2007, 28(6) 1767-75.
[3] Taghavi F, Saghafian H and Kharrazi Y H K. Study on the effect of prolonged mechanical vibration on the grain refinement and density of A356 aluminum alloy[J]. *Materials&Design* 2009 30(5) 1604-11.
[4] Gandin C A, Desbiolles J L, Rappaz M, et al. A three-dimensional cellular automation-finite element model for the prediction of solidification grain structures[J]. *Metallurgical&Materials Transactions* A, 1999, 30(12) 3153-65.
[5] Macro-and Microstructure Evolution of 5CrNiMo Steel Ingots during Electroslag Remelting Process[J]. *Journal of Iron and Steel Research, International* 2014, 21(7) 644-652.
[6] Lashkari O, and Ghomashchi R The implication of rheology in semi-solid metal processes: An overview[J]. *Journal of Materials Processing Technology* 2007, 182(1-3) 229-240.
[7] Kirkwood, D H Semisolid metal processing[J]. *International Materials Reviews* 1994, 39(5):173-189.
[8] Trivedi R and Kurz W. Theory of Microstructural Development During Rapid Solidification[J]. *Acta Metallurgica* 1986, 34(5) 823-830.
[9] Lipton J, Glicksman M E and Kurz W. Dendritic growth into undercooled alloy metals[J]. *Materials Science & Engineering* 1984, 65(1) 57-63.
[10] Wu M, Ludwig A, Bührig-Polaczek A, et al. Influence of convection and grain movement on globular equiaxed solidification [J]. *International Journal of Heat & Mass Transfer* 2003, 46(15) 2819-2832.
[11] Sobczak J J, Drenchev L and Asthana R. Effect of pressure on solidification of metallic materials [J]. *Cast Metals* 2013, 25(1) 1-14.
[12] Spittle J A. Columnar to equiaxed grain transition in as solidified alloys[J]. *Metallurgical Reviews* 2006 51(4) 247-269.
[13] Martorano M A, and Biscuola V B. Predicting the columnar-to-equiaxed transition for a distribution of nucleation under-cooling [J]. *Acta Materialia* 2009, 57(2) 607-615.
[14] Effects of vibration frequency on microstructure, mechanical properties, and fracture behavior of A356 aluminum alloy obtained by expendable pattern shell casting [J]. *The International Journal of Advanced Manufacturing Technology* 2016, 83(1-4) 167-175.

Acknowledgments
We are grateful for grants from the National Natural Science Foundation of China (No. 51775167), and the Qing Lan Project of Jiangsu province.

Fire resistance characteristics of steel-concrete-steel structure and reinforced concrete structure in an eight-lane dual-directional super large-span immersed tunnel

Zhengmao CAO[1, a], **Xiao LIU**[2,b]

[1]China Merchants Chongqing Communications Research and Design Institute Co., Ltd., Chongqing 400067, China
[2]Department of foreign language, Southwest Jiaotong University Hope College, Chengdu 610400, China

E-mail: aswjtuczm@163.com, bxiaoliu13@foxmail.com

Abstract. In order to obtain the fire resistance of Steel-Concrete-Steel structure(SCS) and reinforced concrete structure(RC) in super-large span cross section immersed tunnel, fire numerical simulation calculation was conducted to analyze the temperature field of SCS and RC in an Eight-lane Dual-Directional immersed tunnel fire by the finite element method. Three-dimensional finite element models of steel and concrete composite structure under different tunnel fire scenarios in terms of reasonable thermal parameters, material model, and correct boundary conditions was established. On this basis, analysis for the mechanical properties of the composite steel-concrete structure under the ISO-834 standard fire was conducted. The results of the study indicate that temperature of RC slab near the fire surface is grew faster than SCS slab in the immersed tunnel fire, and the temperature inside the SCS slab is lower than that of RC slab in the immersed tunnel fire. The temperature at the nearest monitoring point by the fire surface of the SCS slab is 31.3% lower than that at the same position in the RC slab, which is 41% lower in the center of SCS slab. Research result provide a reference to the theoretical studies and practical engineering of SCS extra larger span immersed tunnel fire, and provide support for the fire protection design of SCS structures in practical engineering.

Keywords: Tunnel fire, large-span immersed tunnel, steel-concrete-steel structure, numerical simulation, characteristics of fire resistance

1. Introduction

The immersed tunnel is mainly divided into reinforced concrete immersed tunnel, single-layer steel immersed tunnel and double-layer steel reinforced concrete immersed tunnel. At present, apart from Hong Kong, the immersed tunnels built in China are all reinforced concrete structures, the most representative of which is the immersed tunnel of the Hong Kong-Zhuhai-Macao Bridge project. Although the structural type is low in cost, the design method and construction technology are relatively mature, but under the requirements of high water pressure, large section, high durability and uneven settlement, it is still necessary to focus on solving the cracking problem of the reinforced concrete structure, and there is a large limit on the limit net span of the immersed tunnel section.

The steel slab and concrete composite component originally originated in the United States. Compared with the traditional ordinary reinforced concrete technology, there are many advantages such as high rigidity, shock resistance, vibration resistance and fatigue resistance of steel-concrete slab structure, and which is safer. The mechanical behavior of steel-concrete slab structure is more clearly, with better performance, better construction and operation, which is beneficial to shorten the construction period and achieve higher economic efficiency.

In 1976, Solomon proposed a new slab structure with concrete in a double-layer steel slab [1]. Tomlinson proposed to set up shear joints in a double-layer steel-concrete composite structure for the first time, and presented an SCS structural system using a superimposed stud connection [2, 3]. Zeng Jing (2007) of Tsinghua University analyzed the double-plate type composite structure. The research shows that the structure has the advantages of high strength, good ductility, strong impermeability and convenient construction [4]. Wei Fangfang (2012) of Hohai University carried out finite element numerical simulation on the double-steel concrete combined shear wall and proposed a simplified formula for calculating the shear capacity of the composite wall [5]. Yan Jiabao. (2016) of Tianjin University studied the ultimate strength behavior of steel-concrete-steel (SCS) composite plates under concentrated load [6].

Many researches and experiments on SCS composite structures were carried out in the UK, but they were mainly promoted in Japan. Many famous SCS structure tunnels have emerged in Japan, such as Osaka Port Sakishima Tunnel, Kobe Port Minatojima Tunnel, Osaka Port Yumeshima Tunnel and Okinawa Naha Port Tunnel. Fig. 1 illustrates the typical structure of SCS composite structure.

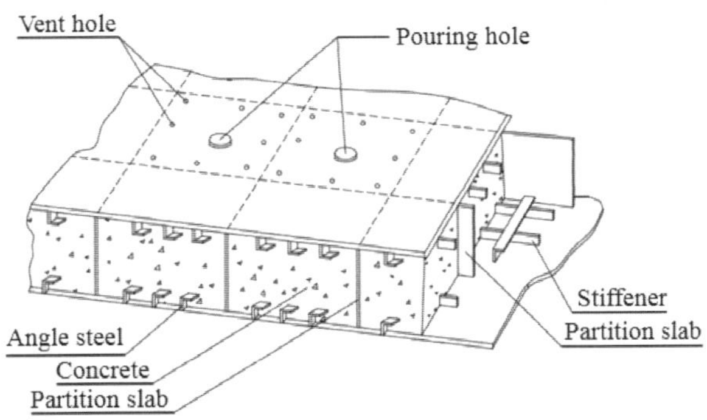

Figure 1. Typical structure of SCS composite structure.

At present, the research on immersed tunnel fires mostly focuses on fires under the reinforced concrete structure. There are some research results in the fire in the steel immersed tunnel [7]. However, there are few studies on the fire in extra-large span eight-lane dual-directional immersed tunnel of SCS structure at home and abroad.

2. Project overview

This work is based on an immersed tunnel project of SCS structure. The tunnel is 6845 meters long, and the cross section is large-span adopting "two tunnels with one pipe gallery" design. The cross section of the whole tunnel is 46m wide and 10.6m high. The thickness of main structure slab is 1.5 m~1.7 m. Fig. 2 illustrates the cross section of the tunnel.

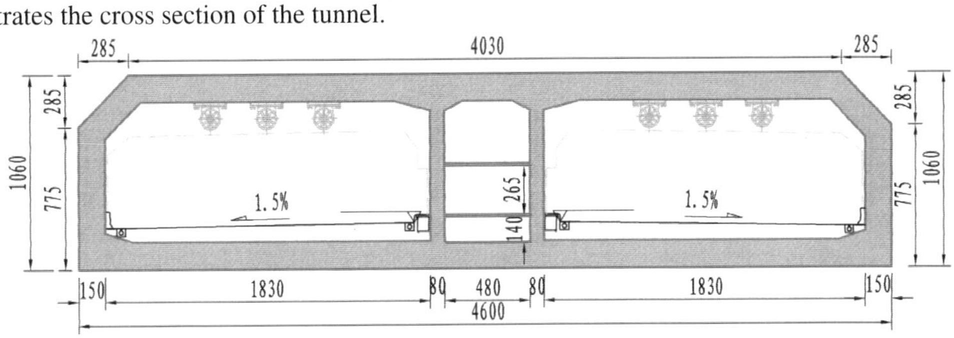

Figure 2. Cross section of eight-lane dual-directional immersed tunnel(cm).

3. Thermal analysis model

The results of multiple researchers are comprehensively compared when selecting the thermal parameters of the material. The thermal parameters of concrete and steel are selected according to the actual design parameters [8]. The effect of water vapor in the concrete is considered when calculating the section temperature field, assuming that the mass percentage of moisture in the concrete is 5%. The specific heat calculation formula of the concrete used is as follows.

$$\rho_c' c_c' = \begin{cases} 0.95\rho_c c_c + 0.55\rho_w c_w & T \le 100^\circ\text{C} \\ \rho_c c_c & T \ge 100^\circ\text{C} \end{cases} \tag{1}$$

$$\rho_w c_w = 4.2\times10^6\ \text{J}/(\text{m}^3\bullet{}^\circ\text{C})$$

where ρ'_c is the mass density of concrete with considering the effect of water vapor; c'_c is the specific heat of concrete with considering the effect of water vapor; ρ_c is the mass density of core concrete without considering the effect of water vapor; c_c is the specific heat of core concrete without considering the effect of water vapor; ρ_w is the mass density of water; c_w is the specific heat of water.

4. Establishment of model

The numerical calculation model of bi-directional eight-lane immersed tunnel is established by fire calculating software. On condition that only a fire happens at the same time. Numerical simulation calculation model is established by actual size of the immersed tunnel of 1:1 along the tunnel cross section. Fig. 3 illustrates the calculation model of an Eight-lane Dual-Directional Immersed tunnel.

The grid size should be 10% of fire characteristic diameter in numerical simulation calculation. Because the area nearby the fire source is sensitive to grid density of numerical simulation, so the density of calculating grid in this area should be increased. The amount of grid in this model is 383,000. Fig. 4 illustrates the calculation model grid setting.

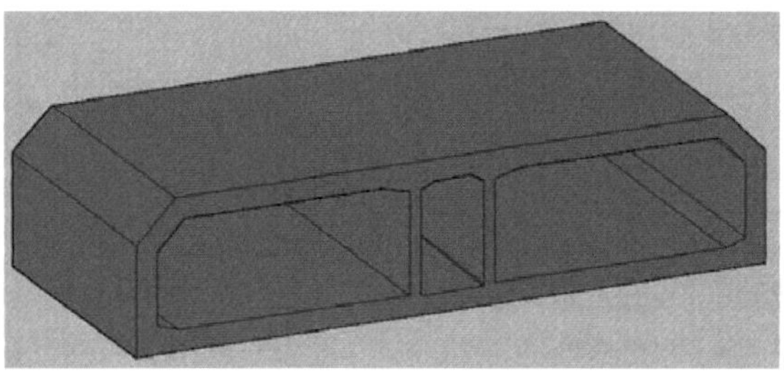

Figure 3. Calculation model of eight-lane dual-directional immersed tunnel.

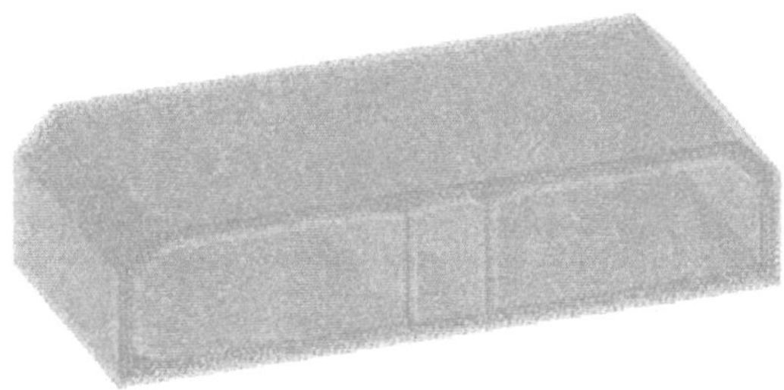

Figure 4. Grid of eight-lane dual-directional immersed tunnel.

5. Parameters setting

Eight-node three-dimensional solid thermal analysis unit DC3D8 is used for steel plates, concrete, support plates and studs when performing finite element modeling, while DC1D2 is for reinforced steel. Temperature changes mainly consider the transfer along the cross-section of the tunnel structure. The transfer along the longitudinal direction of the tunnel is also need to be considered, but not the focus of this paper. Studs and concrete are joined to the steel plate by Tie, the same to steel and concrete. In this way, when the steel-concrete composite slab is fired, the heat can be transferred from the profiled steel to the concrete, the studs and the steel bars. It can effectively simulate the heat transfer inside the component when the actual fire occurs. The heating curve is based on the international standard heating curve ISO-834. Steel-concrete continuous composite floor surface is treated according to the third type of boundary conditions [9]. The boundary condition of fire numerical simulation calculation in immersed tunnel and other settings are depicted in Table 1.

Table 1. Boundary conditions and calculation parameters

Specification	Value	Specification	Value
Heating curve	ISO-384	Stefan-Boltzmann constant	5.67×10^{-8} (m^2•K^4)
Convection coefficient	25W/(m^2•K)	Material of immersed tunnel	SCS/RC
The initial temperature	20 ℃	Width of tunnel single span	18.3 m
Thermal radiation coefficient	0.5	Height of tunnel	6.7 m

6. Result and analysis

Numerical simulation analysis of fire performance was carried out, include SCS immersed tunnel and RC immersed tunnel structure. Figure 5 is a schematic diagram of the arrangement of the temperature monitoring points of the tunnel top slab.

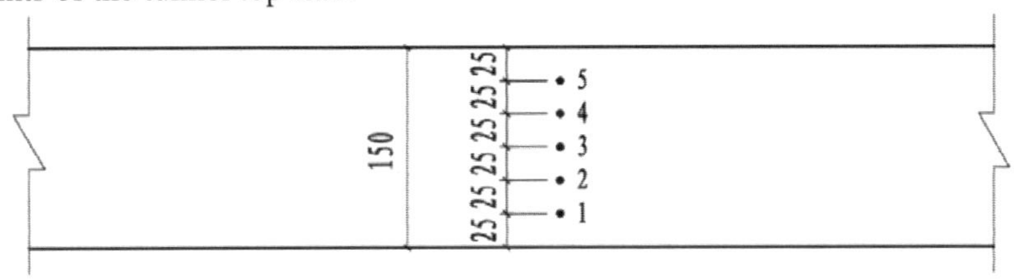

Figure 5. Layout of temperature measuring point of tunnel top slab (units: cm).

Before the numerical simulation calculation, a rectangular section reinforced concrete column model was established and calculated in order to verify the reliability of the calculation model, and compared with the test results obtained by Lie [10]. Figure 6 shows the temperature versus time curve of reinforced concrete columns under fire, where h is the distance from the outer surface of the reinforced concrete column to the inner center point parallel to the central axis of the short side of the rectangular section. It can be seen from the figure that the calculation results are in good agreement with the experimental results, indicating that the established model and the selected calculation method are reasonable and feasible. The temperature of each measuring point in the top slab of the RC immersed tunnel changes with time as shown in Fig. 7. The calculation results of the temperature field of the SCS immersed tunnel structure are shown in Figure 8.

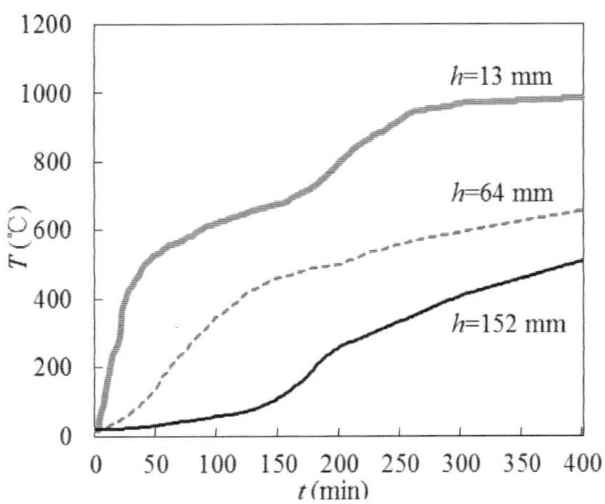

Figure 6. Temperature distribution of reinforced concrete column under fire.

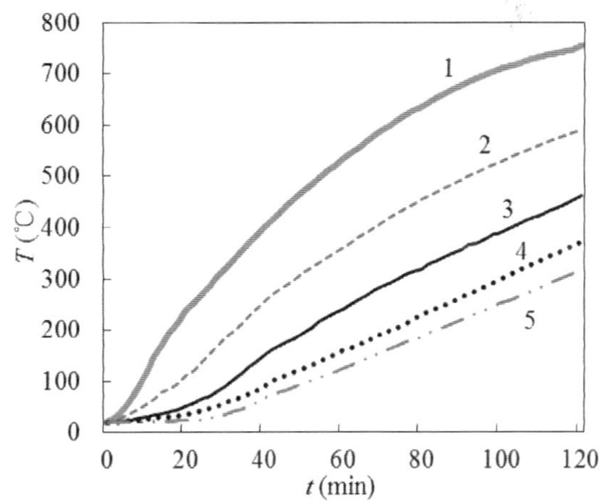

Figure 7. Temperature-time curves of RC immersed tunnel top slab.

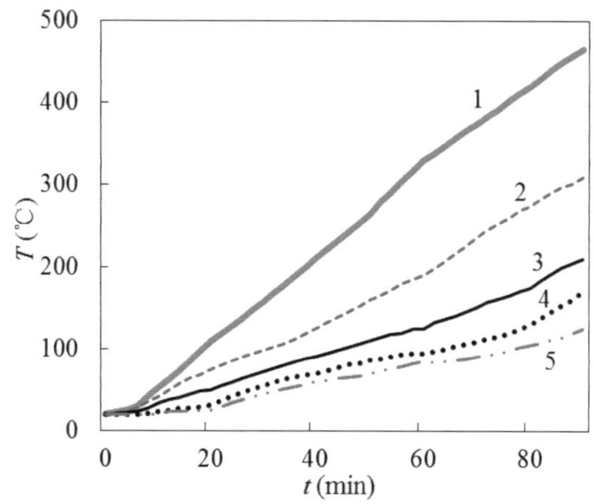

Figure 8. Temperature-time curves of SCS immersed tunnel top slab.

According to the calculation results, temperature field differs greatly at individual points, which is result from the large discrete type of temperature field, and the temperature field heating trend is consistent. The temperature rise of the measuring point close to the fire surface is extremely fast, and the temperature rising speed decreases as the distance between the measuring points and the fire surface increases. At 120 min, the temperature at the measuring point 1 reached 755℃, while the temperature at the measuring point 5 was only 315℃.

It can be seen that the temperature field heating trend is basically the same as the RC structure. Due to the action of the steel plate near the fire surface, most of the heat is absorbed, resulting in a lower temperature in the concrete than in the RC structure. The temperature of RC slab near the fire surface is grew faster than SCS slab in the immersed tunnel fire, and the temperature inside the SCS slab is lower than that of RC slab in the immersed tunnel fire. The temperature at the nearest monitoring point by the fire surface of the SCS slab is 31.3% lower than that at the same position in the RC slab, which is 41% lower in the center of SCS slab.

7. Conclusions

Based on the project of Eight-lane Dual-Directional super-long span immersed tunnel, analysis of temperature field distribution of different material structures in immersed tunnel fire. Temperature field analysis of the SCS immersed tunnel and the RC immersed tunnel was carried out respectively, and the temperature field in the tunnel top slab was compared. The results are as follows:

(1) The calculation model can be used to simulate the temperature field and fire resistance of the steel-concrete composite structure under fire, and it is consistent with the test results. The established model and the calculation method adopted are reasonable and feasible;

(2) For SCS immersed tunnel structure, the temperature in the concrete is lower than that in the RC structure, due to the action of the steel slab near the fire surface, most of the heat is absorbed.

(3) The temperature at the nearest monitoring point by the fire surface of the SCS slab is 31.3% lower than that at the same position in the RC slab, which is 41% lower in the center of SCS slab.

8. References

[1] Solomon S K, Smith D W and Cusens A R *Magazine of Concrete Research* (1976).
[2] M J Tomlinson and A Tomlinson submitted to Immersed Tunnel Techniques (1990).
[3] H Bowerman and J C Chapman *Composite Construction in Steel and Concrete* IV (1998).
[4] J Zeng and J S Fan submitted to *Industrial Construction* (2007).
[5] F F Wei, B Zha and H B Zhao submitted to *Journal of Southeast University* (2012).
[6] J B Yan, J Wang and Y R Liew submitted to *Ocean Engineering* (2016).
[7] S Y Song, J G Nie and G P Xu submitted to *China Civil Engineering Journal* (2013).
[8] J S Han, Y L Dong and W R Chen submitted to *Building Structure* (2006).

[9] G Q Li *Steel-concrete Composite Structure Fire Resistant Design* (China Building Industry Press, China 2006).
[10] B Wu *Mechanical behaviors of reinforced concrete structure after high temperature* (Science Press, China 2003).

Acknowledgments
This paper was supported by "National Key Research and Development Program of China" (Project No. 2018YFC0809600, 2018YFC0809603).

IOP Conf. Series: Materials Science and Engineering **838** (2020) 012011 doi:10.1088/1757-899X/838/1/012011

Analysis of Over-Torque Failure Characteristic of Tool Joints

Feng Chen[1,a], Wei Zhu[1,b], Qinfeng Di[2,c*], Wenchang Wang [2,d], Nan Wang[2,e] and Yuguo Jin[1,f]**

[1]School of Mechatronics Engineering and Automation, Shanghai University, Shanghai 200072, China
[2]Shanghai Institute of Applied Mathematics and Mechanics, Shanghai University, Shanghai 200072, China

Corresponding author：qinfengd@sina.com, wangwenchang1982@163.com

Abstract. Increasingly demanding drilling conditions and progressive drilling technology put forward new requirements for the torsion performance of tool joints. Over-torque failure of tool joints in downhole occurs frequently, resulting in huge economic losses. It is of great significance to analyze the mechanism of over-torque failure of the tool joint. This mechanism can help to selection of the tool joint and improve the design of the tool joint. In this paper, three dimensional elastic-plastic finite element models of single-shoulder tool joint and double-shoulder tool joint are established. Based on ABAQUS, the stress characteristics of tool joints under working torque are analyzed, and the mechanism of over-torque failure of API single-shoulder tool joints and double-shoulder tool joints is revealed. The result shows that there is a serious stress concentration phenomenon in both single-shoulder tool joints and double-shoulder tool joints, which restricts the performance of tool joint. The stress distribution of the tool joint should be homogenized in the design of the new type of tool joint, so as to improve the performance of the tool joint.

1. Introduction

With the decrease of shallow oil resources, drilling and production conditions become more and more harsh, drill string failure accidents often occur. As the weak part of the drill string, tool joints often fail. The cost of a well in the western oilfields of China is about tens of millions of yuan, and some even hundreds of millions of yuan. The wellbore scrap caused by the failure of drill string will cause huge losses [1]. The statistical analysis of drill string failure shows that 65% of the failure accidents occur in the tool joints [2]. Once the failure occurs, the drilling tool will need to be replaced, leading to the delay of the drilling cycle. In serious cases, failure will cause the entire well to be scrapped, causing huge economic losses to the oil field.

In deep wells, ultra-deep wells, and extended reach horizontal wells, the drill string is subjected to a large torque. Especially under extreme working conditions, such as when the drill string gets stuck, it is usually used to move the drill tool by lifting and lowering the drill string. The emergence of top drive equipment makes it possible to rotate the drill string while lifting and lowering the drill tools, which is often accompanied by larger axial tension and rotating torque during operation. Therefore, the tool joint is required to have higher torsion performance [3].

Designing tool joint with higher torsion performance is important for reducing drill string failures. Due to the addition of an secondary shoulder, the double-shoulder tool joint has changed the mechanical characteristics of the structure. Compared with the single-shoulder tool joint in API standard, the double-shoulder tool joint has higher torsional performance and has been widely used. However, due to insufficient understanding of the force mechanism of tool joint, the over-torque failure of tool joint still occurs frequently. By analyzing the over-torque failure mechanism of tool joint, it can help to select the tool joint and improve the design of tool joint.

In this paper, three-dimensional elastic-plastic finite element model of tool joints are established. we use ABAQUS to calculate the stress characteristics of tool joint under the action of torque, and then compare the mechanism of the over-torque failure of the single-shoulder tool joint and the double-shoulder tool joint. The results obtained in calculation are of great significance for improving the design of tool joint.

2. Establishment of three-dimensional finite element model of tool joint

The structure diagrams of single-shoulder tool joint and double-shoulder tool joint are shown in Fig. 1 and Fig. 2. Double-shoulder tool joint has one more shoulder than single-shoulder tool joint. This paper compares the mechanical characteristics of each contact surface of two different types of tool joints, and then the change of torsional performance of tool joints is studied.

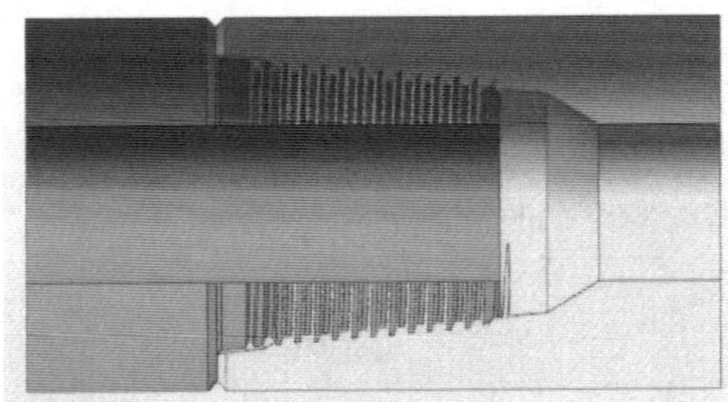

Figure 1. The structure diagrams of single-shoulder tool joint.

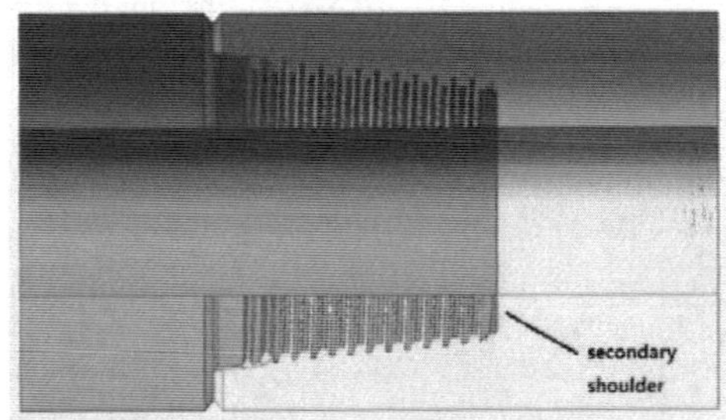

Figure 2. The structure diagrams of double-shoulder tool joint.

The finite element model was established with the 4-3/4″ tool joint as the research object. The basic dimensions of tool joint are shown in Table 1. In order to simulate the stress characteristics of the tool joint under the action of torque accurately, the spiral angle of the thread should be considered and the three-dimensional finite element analysis should be carried out. The three-dimensional finite element models of single-shoulder tool joint and double-shoulder tool joint are established respectively, the meshing situation is shown in Fig. 3.

Table 1. Basic dimensions of tool joint.

Outer diameter [mm]	120.65	Male buckle length [mm]	95.25
Inside diameter [mm]	50.8	Taper	1/6
Base diameter [mm]	89.69	Type	NC35
Female button boring[mm]	98.84		

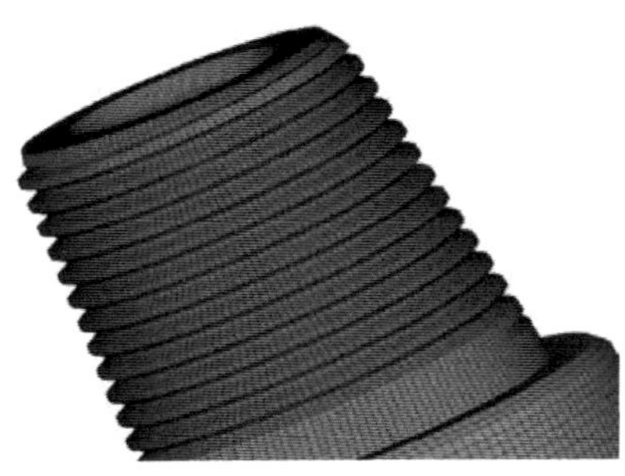

Figure 3. Finite element model of tool joint.

3. Material properties

Performing a tensile test on the material of the tool joint to obtain the relationship between the nominal stress and the nominal strain of the material. According to formulas (1), (2), (3), nominal stress and nominal strain can be transformed into real stress, real strain and plastic strain [4].

$$\boldsymbol{\varepsilon} = \ln(1+\boldsymbol{\varepsilon}_{\text{nom}}) \tag{1}$$

$$\boldsymbol{\sigma} = \boldsymbol{\sigma}_{\text{nom}}(1+\boldsymbol{\varepsilon}_{\text{nom}}) \tag{2}$$

$$\boldsymbol{\varepsilon}^{\text{pl}} = \boldsymbol{\varepsilon}^{t} - \boldsymbol{\sigma} / E \tag{3}$$

where is the real strain, is the real stress, is the nominal strain, is the nominal stress, is the plastic strain, is the total real strain, and is the Young modulus.

Since the test data contains thousands of data points, the elastic-plastic material model should not contain all the data points. Because the force measured by the test has slight up and down fluctuations, the stress-strain curve appears slightly zigzag. This easily leads to the difficulty of calculation convergence [5]. Select the appropriate number of representative data points in the material test data and let it form a smooth stress-strain curve, as shown in Table 2.

According to the test, the plastic strain of the tool joint material failure is 0.192, which is used as the material failure criterion in the three-dimensional elastic-plastic finite element analysis. The text of your paper should be formatted as follows:

Table 2. The relationship between the real stress and the plastic strain.

Real stress [Mpa]	Plastic strain	Real stress [Mpa]	Plastic strain	Real stress [Mpa]	Plastic strain	Real stress [Mpa]	Plastic strain
758	0	920.8	0.016	1017.6	0.045	1079.8	0.075
790.1	0.0005	940.1	0.021	1030.9	0.05	1086.1	0.08
818.3	0.001	953.7	0.025	1043	0.055	1091.2	0.085
878.5	0.006	971.3	0.03	1054.3	0.06	1095.1	0.097
891.3	0.009	987.6	0.035	1063.6	0.065		
905.2	0.012	1003.3	0.04	1072.4	0.07		

4. Three-dimensional Finite Element Analysis of Tool Joint

In order to be closer to the service condition of tool joint, we apply axial tension to the tool joint firstly, and then apply enough working torque to the tool joint until the tool joint fails. Working conditions are shown in Table 3. The loading curve is smooth to avoid numerical oscillation [6].

Table 3. The relationship between the real stress and the plastic strain.

Load of each analysis step	Single-shoulder tool joint	Double-shoulder tool joint
Make-up torque [kN•m]	17.33	26.93
Axial tension[kN]	1500	1500
Torque[kN•m]	40	50

4.1. Single-shoulder tool joint

Under the working torque, the mechanical characteristics of a single-shoulder tool joint are shown in Fig. 4, Fig. 5 and Fig. 6. The calculation results show that when the working torque exceeds a certain value (Referring to Fig. 6, the torque value is approximately 14 kN•m, which is consistent with the conclusion of Brock, J.N. [7]), the pre-tightening state of the tool joint will be broken. This will lead to downhole make-up of too joints, which increases the failure risk of tool joint. In Fig. 4, the contact area on the thread is sharply increased and the contact pressure on the shoulder surface remains substantially unchanged. This is because the threaded teeth are not yet fully contacted under the make-up torque and the axial tension(Especially the middle thread), and the shoulder surface is already in full contact.In Fig. 5, it can be seen that when the working torque exceeds a certain value, the contact force on the primary shoulder and the thread increases sharply, and when it rises to a certain extent, it suddenly drops to zero, this means that the joint of single shoulder drill tool has failed. Fig. 6 shows the working torque loading curve. It can be seen that the applied curve starts to fluctuate when the working torque reaches a certain value. This fluctuation reflects the male and the female engaging surface reaching equilibrium under the action of the working torque, then generating the sliding, then balancing again, then sliding again, and finally destroying. At this point, the material has entered the yield state. Since the yield limit under this condition is relatively stable, the ultimate working torque of the single-shoulder tool joint under this working condition is 24.22kN •m.

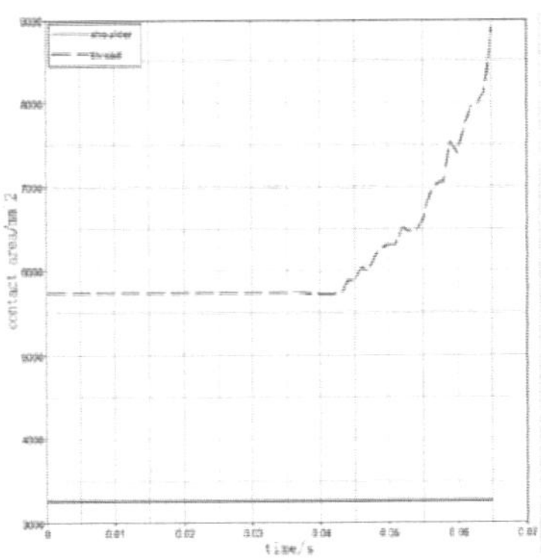

Figure 4. The structure diagrams of single-shoulder tool joint.

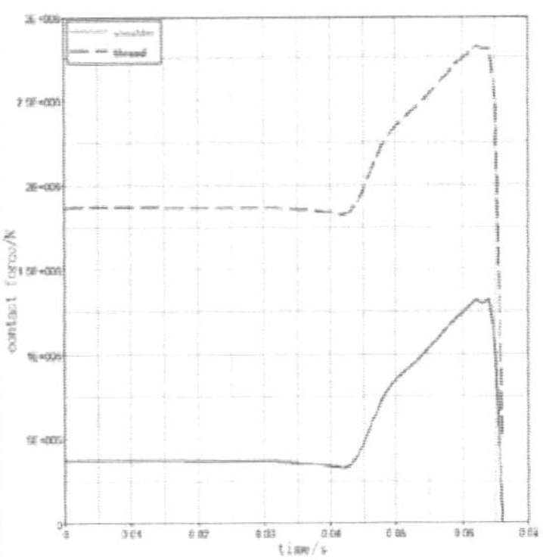

Figure 5. The structure diagrams of single-shoulder tool joint.

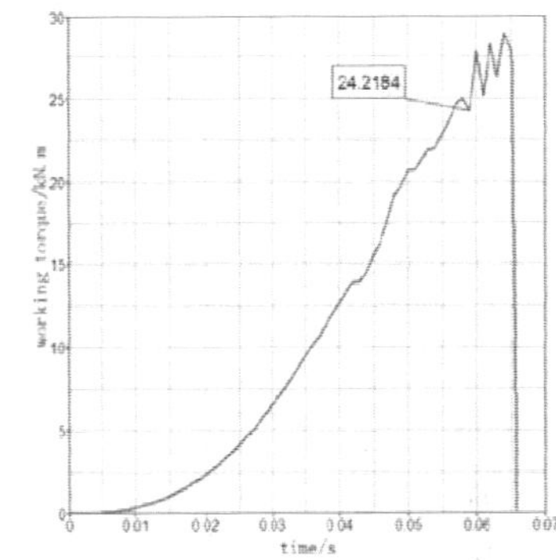

Figure 6. The structure diagrams of single-shoulder tool joint.

The distribution of von Mises stress on the single-shoulder tool joint at the failure time is shown in Fig. 7. It can be seen from the figure that a fault with a zero stress appears at the root of the first fully engaged thread near the shoulder at the time of failure. This is because the stress concentration at the root of the first thread is the most obvious [8]. Compared with other regions, the elements in this region are the first to fail. After the failure of the element, the element is removed, and then the stress concentration is transferred to the adjacent element, gradually expanding, and finally the fracture surface as shown in Fig. 7 appears.

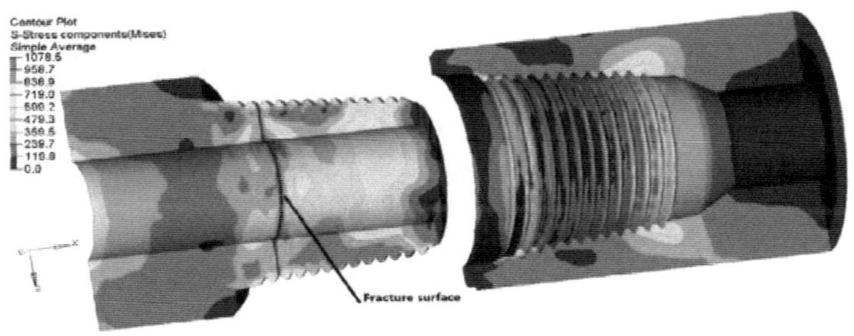

Figure 7. Von Mises stress cloud diagram.

The distribution of equivalent plastic strain on single shoulder tool joint at the failure time is shown in Fig. 8. According to the test, the plastic strain of the tool joint material failed to be 0.192. It can be seen from the figure that at the time of failure, the single shoulder tool joint has a relatively obvious equivalent plastic strain in the front threaded area near the shoulder. This area is a dangerous area, and the field practice is also consistent with this [9].

Figure 8. Equivalent plastic strain cloud diagram.

4.2. Double-shoulder Tool Joint

Under the working torque, the mechanical characteristics of a double-shoulder tool joint are shown in Fig. 9, Fig 10 and Fig 11. Fig. 9 shows that when the working torque exceeds a certain value (referring to Fig. 11, the torque value is approximately 12.5 kN•m), the secondary shoulder that has been separated under the axial tension is resumed contact. It can be seen from Fig. 10 that the contact forces on the primary shoulder, the secondary shoulder and the thread begin to rise sharply. When the working torque continues to be applied to 42.76 kN•m (refer to Fig. 11), it can be seen from Fig. 10 that the contact force on the secondary shoulder suddenly drops to zero, indicating that the secondary shoulder first failed. It can be seen from Fig. 11 that the working torque loading curve also suddenly drops, and then maintains at a certain level and generates fluctuations, eventually failure occurred.

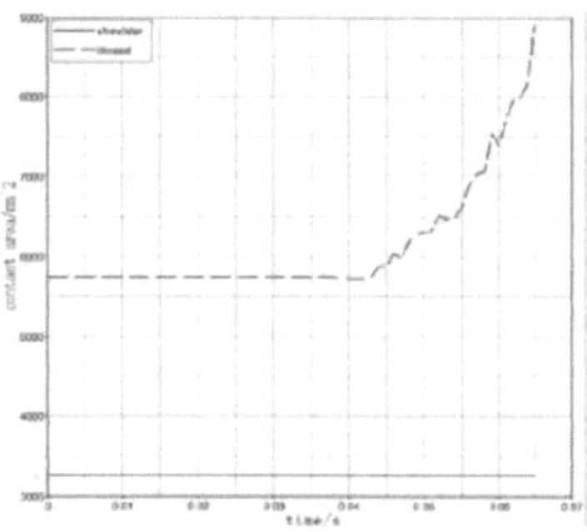

Figure 9. Contact area

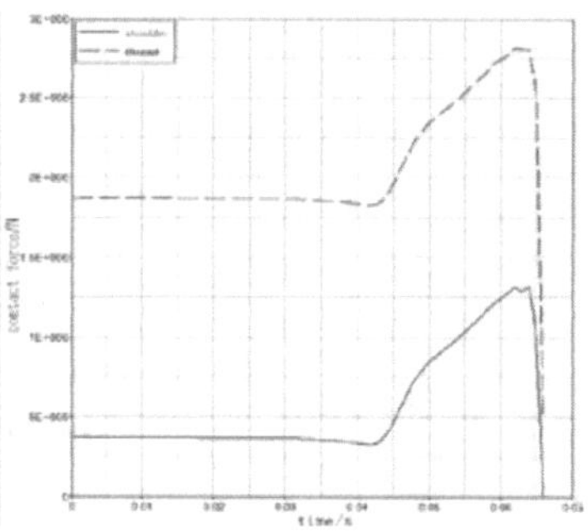

Figure 10. Contact force

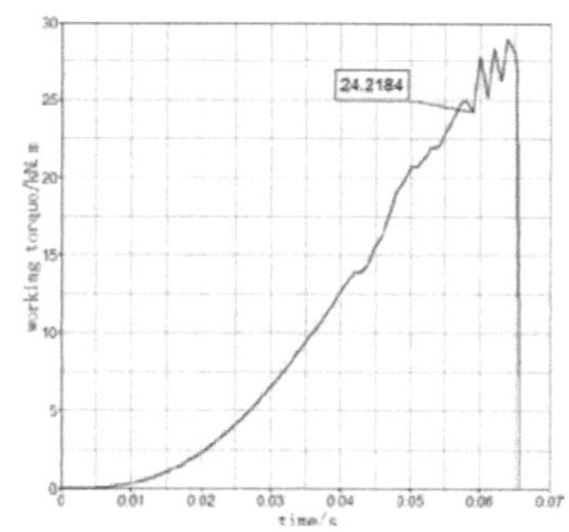

Figure 11. Working torque

The distribution of von Mises stress on the double-shoulder tool joint at the failure time is shown in Fig. 12. It can be seen from the figure that the male fails in the two areas: the area near the root of the first engaging thread and secondary shoulder area. According to Fig. 12, it can be seen that the secondary shoulder has failed first, and finally the root of the first thread has gradually formed a fracture surface. The location of the fracture surface coincides with the joint of a single-shoulder tool joint.

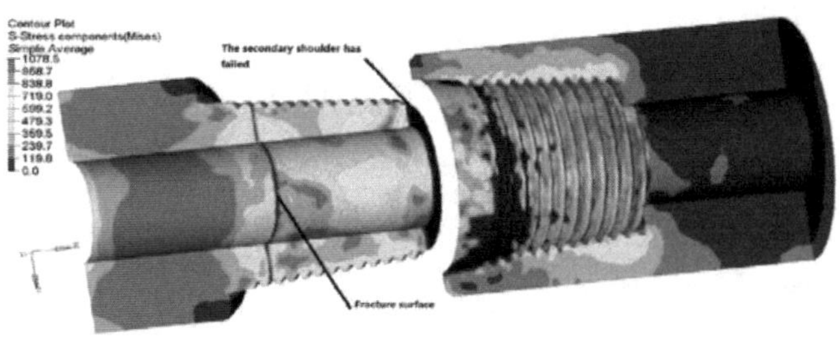

Figure 12. Von Mises stress cloud diagram.

The distribution of equivalent plastic strain on the double-shoulder tool joint at failure time is shown in Fig. 13. It can be seen from the figure that the double-shoulder tool joint has a relatively obvious equivalent plastic strain in two areas at the time of failure. They are the first few threaded areas near the primary shoulder and secondary shoulder area. These two areas are dangerous areas for double-shoulder tool joint, and it is confirmed by field observation [10].

Figure 13. Equivalent plastic strain cloud diagram.

5. Conclusions

(1) Three-dimensional nonlinear finite element analysis can help us understand the stress characteristics of tool joint under complex loads and analyze the failure mechanism of tool joint.

(2) The single-shoulder tool joint has a higher stress in the first few threaded areas near the primary shoulder under the working torque. This area is prone to plastic deformation and is a dangerous area where failure is likely to occur. This single-shoulder tool joint has an ultimate working torque of 24.22 kN•m under this condition.

(3) The double-shoulder tool joint has high stress in the first few threaded areas near the primary shoulder and secondary shoulder area under the working torque. These areas are prone to plastic deformation and are dangerous areas where failure is likely to occur. This double-shoulder tool joint has an ultimate working torque of 42.76 kN•m under this condition, which is 76.5% higher than that of the single-shoulder tool joint.

(4) The results show that there are obvious stress concentration phenomena in the joint of single-shoulder tool joint and double-shoulder tool joint, which is the most fundamental reason restricting the performance of the joint. The stress distribution of tool joints should be homogenized to improve the performance of tool joints.

6. References

[1] Li H, Ji L and Xie L 2006 Current situation analy of oil steel pipe in china *Journal of Hebei University of Science and Technology* pp 1-5

[2] Li B, Yang Z and Yu Z 2006 Finite element analysis of improved design for a drill string joint *Journal of Mechanical Strength* pp 721-4

[3] Chen F, Di Q and Li N 2015 Determination of operating load limits for rotary shouldered connections with three-dimensional finite element analysis *Journal of Petroleum Science and Engineering* pp 622–32

[4] Zhuang Z, You X and Liao J 2009 Finite element analysis and application based on ABAQUS *Tsinghua University Press* p 191

[5] Cao J and Shi Y 2009 ABAQUS finite element analysis faq *Mechanical Industry Press* p 271

[6] Chen F, Di Q and Wang W 2012 The mechanism of the galling for double shoulder drill pipe joint *Chinses Journal of Applied Mechanics*

[7] Brock, James N, Muradov, Andrei, Youngson and Steven S 2011 High breakout torque: hypothesis investigation and evaluation *SPE/IADC Drilling Conference and Exhibition*

[8] Di Q, Chen F and Wang W 2012 Three-dimensional mechanical analysis of the double-shouldered tool joint *ACTA Petrolei Sinica* pp 871-7.

[9] Chen F, Di Q and Di Y 2013 Nonlinear analysis of the threaded connection with three-dimensional finite element model *International Union of Theoretical and Applied Mechanics*
[10] Zhang S, Zhang G and Zhang C 2010 SSC Failure analysis of high strength double shouldered tool joint *Oil Field Equipment*, 2010 pp 71-7.

Acknowledgements
This research was supported in part by the National Science Funding of China (U1663205, 51804194, 51704191), Shanghai Leading Academic Discipline Project (S30106), the Shanghai Municipal Education Commission (Peak Discipline Construction Program), and the Shanghai Sailing Program (17YF1428000).

IOP Conf. Series: Materials Science and Engineering **838** (2020) 012012 doi:10.1088/1757-899X/838/1/012012

Effect of compression deformation on the distortion energy and semi-solid microstructure of Mg-6Al-0.9Gd alloys

W X Wang[1], J Liu[2], J Z Luo[1], S Y Qin[1] and X H Wang[1]

[1]School of Material Science and Engineering, Xi'an University of Technology, Xi'an 710048, China

[2]Faculty of Printing, Packaging Engineering and Digital Media, Xi'an University of Technology, Xi' an 710048, China

Corresponding author's e-mail address: wangwuxiao@163.com

Abstract. In order to investigate the effect of distortion energy on the microstructure of magnesium alloys prepared by strain induced melt activation (SIMA) process and partial remelting, Mg-6Al-0.9Gd magnesium alloys were deformed at 200 ℃ with 0-20% compression ratios, and the relationship between deformation and eutectic melting activation energy was established based on differential scanning calorimetry (DSC) analysis, thereby determining the distortion energy at different compression ratios. The results showed that distortion energy increased with increasing compression ratio, which contributed to the morphological transition from dentrictic to equiaxed structure, and accelerated the formation of liquid phase during subsequent partial remelting. All things (grain size, liquid fraction and distribution in the semi-solid microstructure) considered, the optimal compression ratio for this paper was 20%.

1. Introduction

The Mg-Al series alloys are the most common commercial magnesium alloys due to their good castability and high strength at room temperature, and have received specific attention in the automobile industry [1]. However, their application ranges were restricted due to poor high-temperature strength and creep resistance especially at temperature above 120°C [2]. Previous investigations indicated that addition of rare earth elements (RE, e.g. Gd, Y, Nd, Sm, Dy) to Mg-Al alloys can effectively improve their high temperature strength and creep resistance by generating thermally stable precipitates $Al_{11}RE_3$ or Al_2RE phases and suppressing the formation of β-$Mg_{17}Al_{12}$ phases [3].

To date, almost 90% of total magnesium alloy components are produced by casting processes, especially high pressure die casting (HPDC) [4]. In recent years, semi-solid processing had become popular for industrial and commercial production, and exhibited significant advantages over HPDC, such as longer die life due to lower temperature, better mechanical properties due to less entrapped air, less porosity and fine microstructure, and near-net shape processing [5]. It is well known that the key of semi-solid processing is to prepare semi-solid slurry with a non-dendritic, spheroidal microstructure. Many methods such as electromagnetic stirring [6], gas bubbling process [7], recrystallization and partial melting [8], and strain induced melt activation (SIMA) have been used to prepare semi-solid slurry [9]. In the SIMA process, the billet with dendritic crystal is pre-deformed (compression, forging, extrusion, e.t.), and then reheated to the semi-solid state temperature followed by holding isothermally to make the dendritic crystal sphericized. Finally, the semi-solid billet is formed into required component. Through the pre-deformation, a certain number of distortion energy accumulates in the deformed billet, which contributes to the microstructure transition from dendritic crystal to equiaxed grains at low temperature in a short time, preventing microstructure coarseness during holding process at elevated temperature [10]. Therefore, it is of great significance to study the correlation between distortion energy and the semi-solid microstructure of the deformed metals. However, there are only limited amount of reports about the quantitative examination of distortion energy, because it cannot be experimentally measured directly [11]. Considering the eutectic melting activation energy when atom transits from solid phase to liquid phase during melting can be determined by calculation [12], it is necessary to investigate the relationship between distortion energy and eutectic melting activation energy. In present work, Mg-6Al-0.9Gd alloys were subjected to hot compression deformation, and the microstructure of the compressed specimens with different compression ratios was observed. The relationship between eutectic melting activation energy and distortion energy was examined through DSC analysis, thereby determining the accumulated distortion energy at different compression ratios. The effect of distortion energy on the semi-solid microstructure was discussed. This work is expected to provide theoretical guidance in semi-solid processing of Mg alloys.

2. Experimental procedures

High-purity Mg (99.95 wt. % purity), high-purity Al (99.98 wt. % purity), Mg-30%Gd master alloy were used as raw materials to prepare Mg-6Al-0.9Gd magnesium alloys. Cylindrical specimens with the same volume but different dimensions (ø25.2mm×61.3mm, ø24.5mm×64.8mm, ø23.8mm×68.7mm) were machined from the obtained ingot casting. These specimens were heated to 200 ℃ and held for 60 min, and then compressed into ø26.6 mm×55mm cylindrical specimens using closed-die forging in a WE-600 universal materials tester. The compression ratio of the specimens was defined as $\varepsilon = (H_0 - H)/H_0$, where H_0 and H are specimen height before and after compression deformation, respectively. Hence, the compression ratios of the three cylindrical specimens are 10%, 15%, and 20%, respectively.

Both the samples for metallographic observation and DSC tests were machined from the section center of the compressed specimens. In order to examine the effect of compression ratio on the solidus and liquidus temperatures, DSC tests were performed on a NETZSCH STA 449F3 thermal analyzer under high-purify argon gas atmosphere. The samples with 3 mm in diameter and about 30 mg in weight were heated to 800 ℃ at 5 K/min, 10K/min, and 15K/min, respectively, and then cooled to room temperature. In order to study the effect of compression deformation on the semi-solid microstructure, the compressed specimens were subjected to a isothermal heat treatment at 550 ℃ for 15 min, and subsequently quenched in cold water. The metallographic specimens were mechanically ground, polished and etched with a 4 vol. % nitric acid-ethanol solution, and then examined by using an Olympus GX71 optical microscope (OM). Grain size ($D = \sqrt{4A/\pi}$, where A is the area of solid grain), and liquid phase fraction were measured from the resulting optical micrographs using an image analysis system.

3. Results and discussion

3.1. Deformed microstructure

Figure1 shows the microstructure of Mg-6Al-0.9Gd alloys at different deformation ratios. Figure 1 (a) shows the original cast microstructure, which is characterized by randomly distributed coarse dendritic crystal. Figure 1 (b) shows the microstructure of the compressed specimen at 10% compression ratio. It can be seen from Figure 1 (b) that the microstructure still presents dendritic morphology, and only partial dendrites were broken. When compression ratio reaches 15%, some dendrite arms were broken into fragments, some dendrite arms become thinner and longer due to compression deformation, and the space between adjacent dendrite arms decreased obviously, as shown in Figure 1 (c). When compression ratio rises to 20%, the number of fractured dendrites further increased, and finer particles were formed in some areas, as shown in Figure 1 (d). During compression deformation, some large dendritic crystals were squashed, bent, and broken into fine grains. The grains rearranged and rotated under flow stress, forming wavy stripes and causing plastic deformation of the whole specimen.

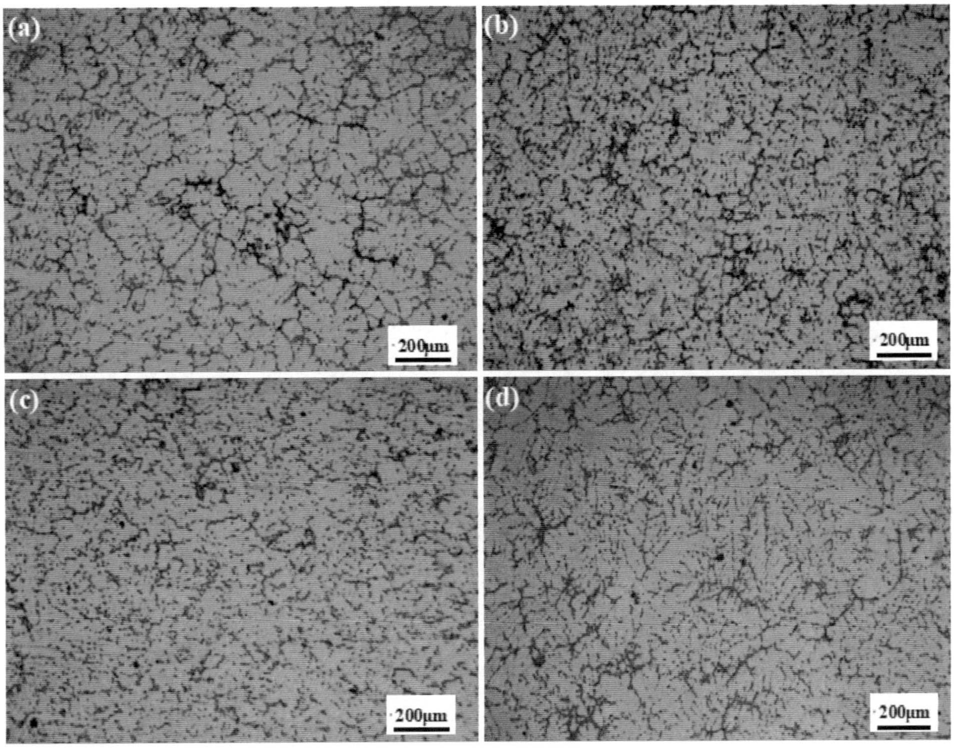

Figure 1. Optical microstructure of Mg-6Al-0.9Gd alloys at different compression ratio (a) 0%, (b) 10%, (c) 15%, and (d) 20%.

The microstructure of the Mg-6Al-0.9Gd alloys gradually changed with increasing compression ratio can be ascribed to following aspects: on the one hand, at the early stage of compression deformation, some dendrite arms were stretched due to plastic deformation, which cause the increase of intragranular defects. On the other hand, the dendritic crystals which are not easy to be deformed plastically were fractured, which introduced intergranular defects. The accumulation of micro defects resulted in the increase of distortion energy.

The difference in morphology of dendritic crystals became greater and greater with increasing compression ratio, and this can be due to the difference between α-Mg dendritic crystals and β-$Mg_{17}Al_{12}$ phase. The former are soft phases, and the latter are hard phases at grain boundary. When deformation was small, the soft dendritic crystals underwent slight deformation. With the increase of deformation, the soft dendritic crystals were stretched, and suffered from obvious plastic deformation. However, the hard β-$Mg_{17}Al_{12}$ phases which are not easy to be deformed plastically were ruptured.

3.2. Eutectic melting activation energy

The compression deformation caused the accumulation of distortion energy in the deformed Mg-6Al-0.9Gd alloys. When eutectic reaction occurs during subsequent partial remelting, the thermodynamic behavior of the deformed samples will vary with different deformation. Figure 2 shows the DSC curves obtained at a heating rate of 10 K/min for the deformed Mg-6Al-0.9Gd alloys with different compression ratios, in which T_p is peak melting temperature, i.e. the temperature corresponding to the maximum deflection of the DSC curve, and Ton is onset melting temperature, i.e. the temperature found by extrapolating the baseline (prior to the peak) and the tangent at the inflection point on the leading side of the peak to their intersection [13], as shown in Figure 2 (d). It can be seen that all the endothermic peaks in the DSC curves are sharp with narrow temperature range. Combining the microstructure of the deformed Mg-6Al-0.9Gd alloys and phase diagram [14], it can be inferred that these endothermic peaks correspond to eutectic reaction, i.e. α-Mg+ β-$Mg_{17}Al_{12}$→L.

Onset melting temperature T_{on} refers to the minimum temperature at which endothermic effect occurs, and is one of the most important data in DSC analysis. Table 1 lists the onset melting temperatures determined from DSC curves. Onset melting takes place at the early stage of eutectic structure transformation. From Table 1, it can be found that at the same heating rate, Ton decrease somewhat with increasing compression ratio. Under the same compression ratio, Ton increases with increasing heating rate.

The peak melting temperatures T_p under different conditions are shown in Table 2. From Table 2, it can be noticed that the peak melting temperature of the original material is 710.14 K when heating rate is 10K/min. As the compression ratio increased from 10% to 20%, the peak temperature decreased from 709.96 K to 709.37 K, i.e. at the same heating rate, T_p decreases with increasing compression ratio. This is because with the increase of compression ratio,the coarse dendritic crystals were squashed, bent, and broken into fine particles, and the whole specimen was deformed, releasing heat. The larger the compression ratio is, the more the released heat is. Hence, the required heat of the deformed alloys during remelting decreased, i.e. T_p decreased consequently. Under the same compression ratio, T_p increases with increasing heating rate. This is due to that with the increase of heating rate, generated heat flow per unit time increases, T_p becomes higher and higher consequently. In addition, the atoms in the alloys move faster, thermal inertia increases, energy fluctuates, hence T_p increase.

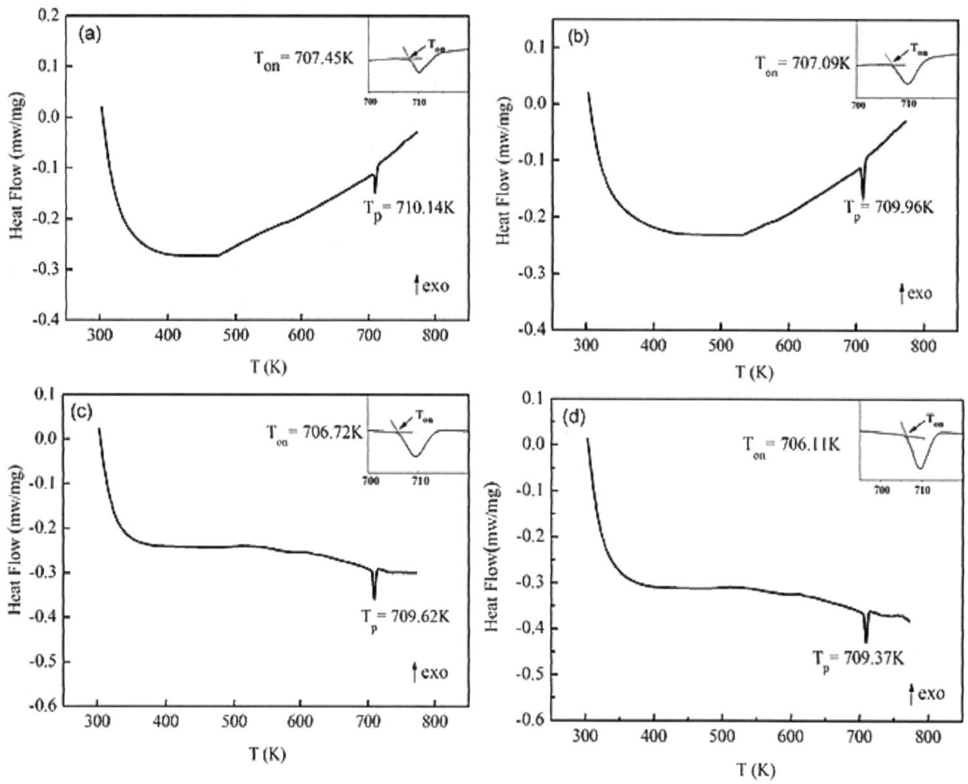

Figure 2. DSC curves of Mg-6Al-0.9Gd alloys with different compression ratio (a) 0%, (b) 10%, (c) 15%, and (d) 20%, heating rate is 10K/min.

Eutectic melting activation energy can be determined using Kissinger formula [15],

$$d\left[\ln\left(\varphi/T_P^{\ 2}\right)\right]/d\left(1/T_P\right) = -E/R \tag{1}$$

where φ is heating rate (K/min), T_p is the peak melting temperature obtained from differential thermal analysis, E is eutectic melting activation energy (kJ/mol), and R is gas constant (8.314 kJ/mol). Figure 3 shows the relationship between $\ln(\varphi/T_P^2)$ and $1/T_p$, which is fairly linear. The eutectic melting activation energy E can be determined by calculating the slope of the fitted straight lines according to (1). Figure 4 shows the eutectic melting activation energy of the deformed Mg-6Al-0.9Gd alloys under different compression ratios.

Table 1. Ton of the Mg-6Al-0.9Gd alloys at different compression ratio, (K).

Heating rate (K/min)	Compression ratio (%) 0	10	15	20
5	707.33	706.90	706.52	705.83
10	707.45	707.09	706.72	706.11
15	707.88	707.24	706.91	706.28

Table 2. Tp of the Mg-6Al-0.9Gd alloys at different compression ratio, (K).

Heating rate (K/min)	Compression ratio (%) 0	10	15	20
5	709.51	709.37	709.21	708.99
10	710.14	709.96	709.62	709.37
15	710.92	710.77	710.37	710.05

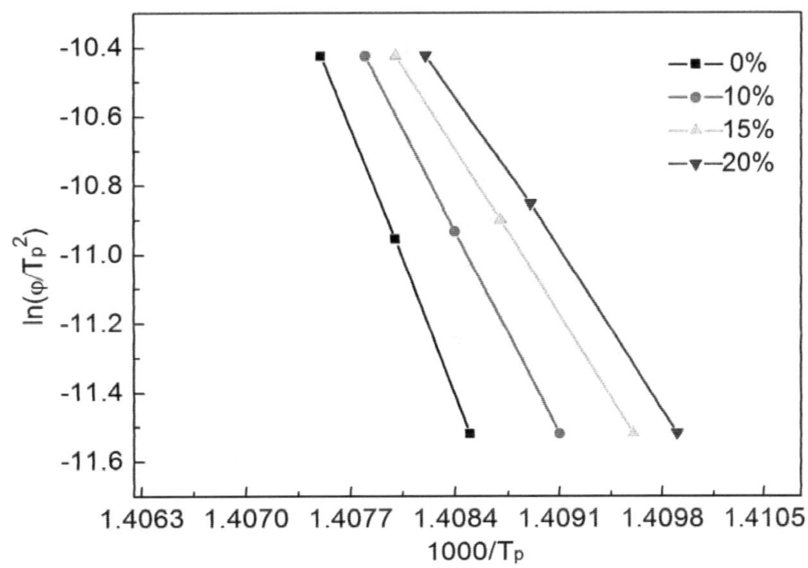

Figure 3. The relationship between $\ln(\varphi/T_P^2)$ and the $1/T_p$.

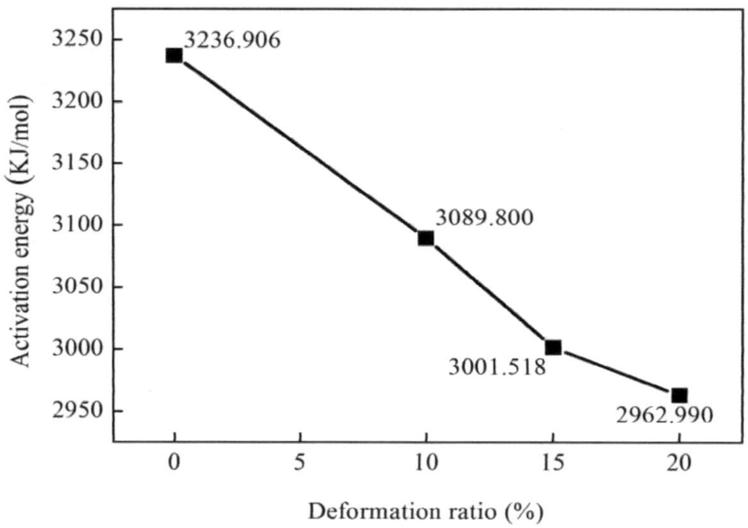

Figure 4. Eutectic melting activation energy at different compression ratios.

When compression ratio increases from 0 to 20%, the eutectic melting activation energy of the Mg-6Al-0.9Gd alloys decreases from 3236.91 kJ/mol to 2963.10 kJ/mol. This is due to that with the increase of compression ratio, the density of serious defects (e.g. grain boundary, dislocation, and vacancy) gradually increases, the disorder degree of crystal structure increases consequently, which causes the reduction of eutectic melting activation energy.

The distortion energy stored in the deformed sample can be regarded as the difference between the eutectic melting activation energy of the sample before and after deformation [16]. Table 3 shows the obtained distortion energy based on eutectic melting activation energy corresponding to different compression ratios. It can be found that with the increase of compression ratio, eutectic melting activation energy decreases from 3236.91 kJ/mol to 2962.99 kJ/mol, while the distortion energy accumulated to 273.92 kJ/mol after 20% deformation. In other word, partial eutectic melting activation energy transformed into distortion energy stored in the deformed alloys.

When compression ratio is 10%, the distortion energy stored in the alloys is very little, and the dendritic crystals have the tendency of fracture, but no obvious deformation has taken place. When compression ratio reaches 15%, the dendrite arms become long and thin due to plastic deformation, and the density of the generated defects increases, causing the distortion energy to increase. When compression ratio rises to 20%, the number of fractured dendrites increases, which is a process of stress relief, causing the distortion energy to decrease. However, the soft dendrite arms were further stretched and become longer and thinner, causing the distortion energy to increase consequently. The variation of distortion energy is consistent with the evolution of microstructure.

Table 3. The activation energy and distortion energy of the deformed Mg-6Al-0.9Gd alloys at different compression ratios.

Compression ratio(%)	0%	10%	15%	20%
Activation energy(kJ/mol)	3236.91	3089.80	3001.52	2962.99
Distortion energy(kJ/mol)	0	147.11	235.39	273.92

3.3 Effect of deformation on semi-solid microstructure

Figure 5 shows the microstructure of Mg-6Al-0.9Gd alloy at different compression ratios with isothermal holding temperature of 550 °C for 15 min. It is found that the microstructure of the compressed alloys has changed into equiaxed grains surrounded by liquid phase at grain boundary Under compression ratio 10%, some dendrite arms were melted and the liquid phase precipitated at grain boundaries, mainly segregated at triple junctions, as shown in Figure 5 (a). Then the grain boundaries were irregular and chaotic due to the incipient melting of eutectic phase there. Moreover, there exist some small entrapped liquid pool inside the solid particles. When the compression ratio increases to 15%, the liquid pool inside some small grains connected with the liquid phase at grain boundaries, forming network of liquid phase as shown in Figure 5 (b). When the compression ratio reaches 20%, the amount of liquid phase increased obviously and penetrated most grain boundaries as shown in Figure 5 (c), and it is beneficial for improving the liquidity of the alloy during semi-solid processing.

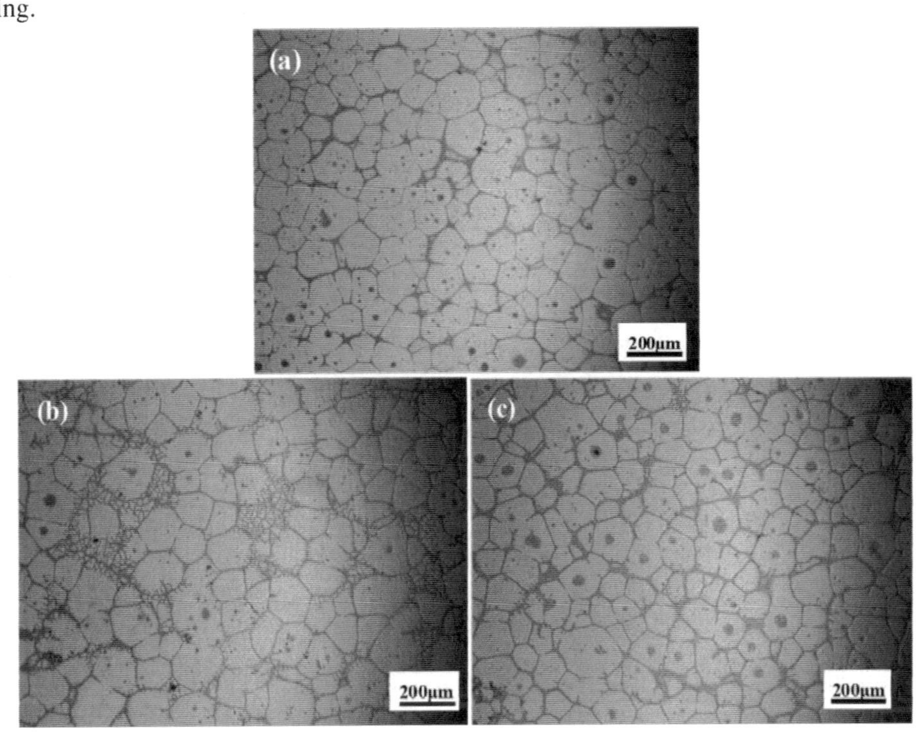

Figure 5. Optical micrographs of the deformed alloys at different compression ratios after isothermal heat treatment at 550 °C for 15 min, (a) 10%, (b) 15%, and (c) 20%.

Figure 6 shows the variations of liquid fraction and average grain size with compression ratio. With the compression ratio varying from 10 % to 20 %, the average grain diameter decreases from 55.08 μm to 49.11 μm, and the liquid fraction increases from 13.19 vol.% to 29.86 vol.%. During the compression deformation, distortion energy was stored in the alloys, and the energy accumulated with increasing compression ratio. It is well known that the distortion energy stored in the alloys provide driving force for recrystallization, which can be verified by the occurrence of many strain-free equiaxed grains as shown in Figure 6. The initial recrystallized grain size during partial remelting depends mainly on the amount of stored energy, and decreases with increasing distortion energy. The smaller the grain size was, the more the grain boundary area was, thus more melting paths were provided for the penetration of liquid phase. When the remelting temperature rises to above solidus, the deformed alloys with more grain boundary have more solid-liquid interfacial energy, and thus have higher diffusion rate. Consequently, the formation rate of liquid phase in the compressed alloys was enhanced, and the liquid fraction in the alloys increased with increasing compression ratio under the same condition.

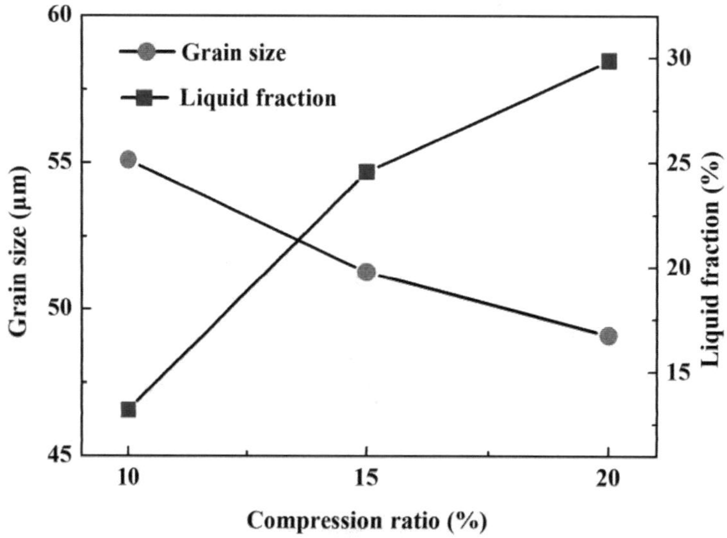

Figure 6. The variations of average grain size and liquid fraction in the semi-solid alloys.

4. Conclusions

1) The distortion energy stored in the deformed Mg-6Al-0.9Gd alloys increased with increasing compression ratio. When compression ratio increased from 10% to 20%, the distortion energy increased from 147.11 kJ/mol to 273.92 kJ/mol.
2) With the increase of deformation, the defect density increased, and the distortion energy was stored in the deformed Mg-6Al-0.9Gd alloys in forms of dislocation multiplication.
3) The optimum compression ratio should be 20%, at which the average diameter of equiaxed grain reached the minimum, and the fraction of liquid phase raised to the maximum after a isothermal treatment at 550 ℃ for 15 min.

5. References

[1] William J J and Krajewski P E 2017 *Scripta Mater.* **128** 107-12
[2] Liu J, Wang W X, Zhang S, Zhang D J and Zhang H Y 2015 *J. Alloys Compd.* **620** 74-9
[3] Wang Z W, Yan H, and Huang W X 2016 *China foundry* **13** 383-88
[4] Luo A A 2013 *J. Magnes. Alloy* **1** 2-22
[5] Zhang Y,Li X P, Sun S P, Lu Y L, and Wu G H 2017 *China foundry* **14** 485-91
[6] Y. S. Chen, L. Zhang, W. C. Liu, G. H. Wu, and W. J. Ding 2016 *Materials Design* **95** 398-409
[7] Zhang Y, Wu G H, Liu W C, Zhang L, Pang S and Ding W J 2015 *Transactions of Nonferrous Metals Society of China* **25** 2181-87
[8] Chen Q, Zhao Z D, Chen G and Wang B 2015 *J. Alloys Compd.* **632** 190-200
[9] Wang L P, Jiang W Y, Chen T, Feng Y C, Zhou H Y, Zhao S C, Liang Z Q and Zhu Y 2012 *Transactions of Nonferrous Metals Society of China* **22** 435-44
[10] Jiang J F, Wang Y, Liu J, Qu J J, Du Z M and Luo S J 2013 *Transactions of Nonferrous Metals Society of China* **23** 576-85
[11] Cao M, Zhang Q and Zhang Y S 2017 *J. Alloys Compd.* **721** 220-28
[12] Shi C J and Chen X G 2016 *Mater. Sci. Eng.* A **650** 197-209
[13] Chen S Y, Chang G W, Huang Y D and Liu S Y 2013 *Transactions of Nonferrous Metals Society of China* **23** 1602-09
[14] Risueño E, Doppiu S, Rodríguez-Aseguinolaza J, Blanco P, Gil A, Tello M, Faik A and D'Aguanno B 2016 *J. Magnes. Alloy* **685** 724-32
[15] Sanaty-zadeh A, Xia X, Luo A A and Stone D S 2014 *J. Magnes. Alloy* **583** 434-40
[16] Wang W X and Xing J D 2008 *Rare metal materials and engineering* **37** 354-56 (In Chinese)

Acknowledgements
This work was supported by the National Nature Science Foundation of China (Grant no. 51771146) and Natural Science Basic Research Plan in Shaanxi Province of China (Grant no. 2019JM-180).

IOP Conf. Series: Materials Science and Engineering **838** (2020) 012013 doi:10.1088/1757-899X/838/1/012013

Evolution of microstructure and toughness of very tough ferritic stainless steels

F Gao[1,2], F Liu[1,2], C S Fang[1], W Zhang[3], J J Zhang[4] and Y Zou[4]

[1]School of Materials Science and Engineering, Northeastern University, Shenyang 110819, Liaoning Province, China
[2]Key Laborotary of Crucial Lightweight Structural Metallic Materials, Liaoning Province, Northeastern University, Shenyang 110819, Liaoning Province, China
[3]CITIC-CBMM Microalloying Technology Center, CITIC Metal Co., Ltd., Beijing 100004, China
[4]Technology Center, Taiyuan Iron & Steel (Group) Co., Ltd., Taiyuan 030003, Shanxi Province, China

E-mail: gaof@mail.neu.edu.cn

Abstract. In present work, the hot rolling with different finish rolling temperatures and annealing with different annealing temperature are employed with high toughness AISI444 ferritic stainless steel, and different microstructure are generated. It was indicated that low finish rolling temperature could lead to the significant refinement of the rolled and annealed microstructures and decrease the fraction of grains with {001} planes which are parallel to notch plane in the annealed microstructure. Therefore, the ductile-brittle transition temperature was lowered from -44.2°C to -48.8°C, resulting in the improvement of toughness after employing the low finish rolling temperature for ferritic stainless steel. When high annealing temperature can be applied, the coarsening of annealed microstructure occurred, but the fraction of grains with {001} planes which are parallel to notch plane in the recrystallized microstructure decreased. Thus, the ductile-brittle transition temperature has no significant change after annealing with different temperatures.

1. Introduction

Ferritic stainless steels (FSSs) can replace the conventional Cr-Ni austenitic stainless steels (ASSs) for specific applications, such as heat exchanger shell, because of the face that FSSs has better stress corrosion cracking resistance and lower cost than ASSs [1,2].

However, as compared with ASSs, FSSs show high ductile-brittle transition temperature (DBTT), especially when their thickness is greater than 5mm because of specimen size effect which means that thicker specimens possess higher DDBT [2]. This poor toughness can results in the cracking during stamping and the significantly decrease of toughness after welding for FSSs medium plate, which will impedes the extensive application of FSSs. Our previous work [1] suggest that the occurrence of ductile-to-brittle transition can be closely connected with the formation of deformation twin: during Charpy impact testing, deformation twins could form in FSSs; a microcrack tend to initiate at the tip of deformation twins and propagate along the boundary of twins, leading to the occurrence of brittle fracture. This suggests that parameters which can impact the occurrence of deformation twinning will influence the DBTT and toughness of FSSs [1]. Based on this work, the alloy design of high toughness

FSSs is developed [2] in terms of refining microstructure and increasing stacking fault energy which can suppress twinning.

In present work, high toughness AISI444 FSS are hot rolled and annealed by different processes for further decreasing the DBTT and optimizing the notch toughness. The effects of hot rolling and annealing on microstructure and toughness were investigated.

2. Experimental procedures

A high toughness AISI444 FSS used in this study has the composition of 0.008% C, 0.0041% N, 0.16% Nb, 0.8% Ni, 0.43% Cu, 0.06% Al, 1.8% Mo, 18.5% Cr, 0.05% Si, 0.01% Mn and the balance Fe. After a homogenization treatment of 2h at 1200℃, the cast ingot was rough rolled at temperature of 1100-1050°C. Subsequently, one of the rough rolled plates was cooled in air to 1030℃ and then finish rolled with the finish rolling temperature of 914°C. The other rough rolled plate was cooled in air to 905°C and then finish rolled with the finish rolling temperature of 854°C. The former is denoted as HFT (High finish rolling temperature), and the latter is referred to as LFT (Low finish rolling temperature). After different hot rolling processes, an annealing treatment of 8 min at 1000°C was undertaken for the hot rolled plates HFT and LFT for study the influence of hot rolling conditions on microstructure and property, and the plates LFT were annealed at temperature of 900-1000°C for 8 min for study the influence of annealing conditions on microstructure and property.

The microstructure and microtexture were observed with the optical microscope (OM) and the Electron Backscatter Diffraction (EBSD) system, respectively. The subsize Charpy specimens were machined along the rolling direction (RD) for Charpy impact tests at temperatures of -80-20°C.

3. Results and discussion

Figure 1 shows the toughness of annealed plates with LFT and HFT processing routes and the DBTT vs. plate thickness for AISI444 FSSs. The DBTTs were -44.2 and -48.8°C for the annealed plates HFT and LFT, respectively. This indicates that decreasing finish rolling temperature leads to the decrease of the DBTT. The DBTTs were -50.0, -51.0 and -48.8°C for the annealed plates annealed at 900, 950 and 1000°C, respectively, indicating that the annealing temperature has no obvious influence on the DBTT. AISI444 FSSs medium plate developed in this study has more superior toughness than those using conventional alloy design and processing routes.

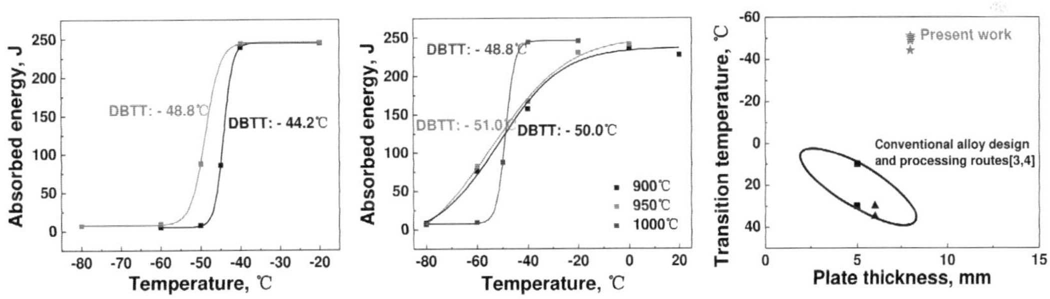

Figure 1. The toughness for annealed plates with different rolling (a) and annealing (b) processes and the DBTT vs. plate thickness for AISI444 FSSs obtained in present work and reported by other researcher using conventional alloy design and processing routes [3,4] (c)

Figure 2 shows the rolled microstructures observed by OM and recrystallized microstructures analyzed by EBSD after different rolling conditions. Low finish rolling temperature can result in larger accumulated deformation stored energy and lower dynamic softening rate [1], so the hot rolled plate LFT possesses higher dislocation density, more in-grain shear band and finer microstructure than rolled plate HFT, as shown in Figures 2(a) and (c). In the annealed plate HFT, the partial recrystallization occurred and the average grain size is found to be about 240μm obtained by EBSD analysis (Figure 2(b)). By contrast, full recrystallization was observed, the microstructures are consisted of equi-axed grains in annealed plate LHT and the average grain size is found to be about

107μm obtained by EBSD analysis (Figure 2(d)). The annealed plate LHT has finer and more homogeneous microstructure than the annealed plate HF, as shown in Figures 2(b) and (d). This can be due to high dislocation density and fine microstructure in the hot rolled plate LHT, which can increase the recrystallization driving force and nucleation sites of recrystallized grain [3].

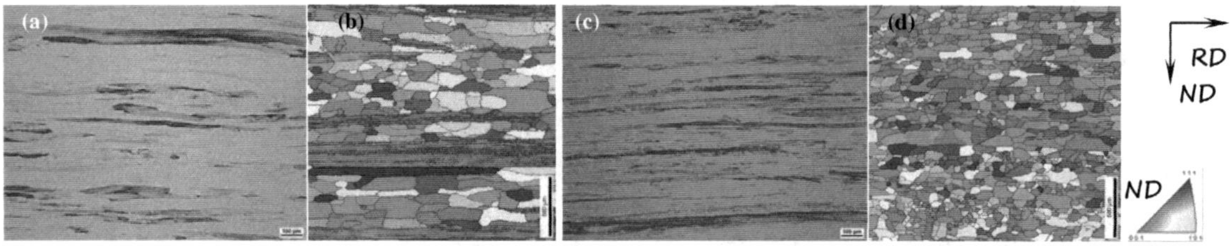

Figure 2. The microstructures of the rolled plates ((a),(c)) and orientation maps of all orientations of annealed plates ((b),(d)) with different rolling conditions (a),(b) HFT, (c),(d) LFT

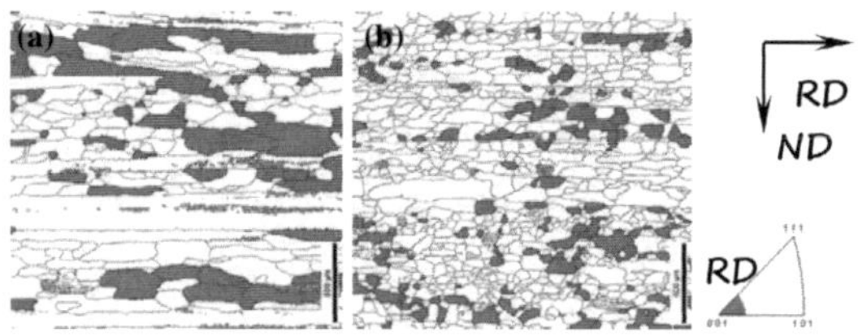

Figure 3. The orientation maps of <001>//RD oriented grains measured from the annealed plates with different rolling conditions (a) HFT, (b) LFT

Figure 3 shows the orientation maps of <001>//RD oriented grains measured from the annealed plates with different rolling conditions. The fraction of <001>//RD oriented grains in the annealed plates HFT and LFT are measured to be 25.5% and 17.1%, respectively, indicating that after low finish rolling temperature the fraction of the <001>//RD oriented grains markedly decreases. Hot deformation at lower temperature can promote the crystal rotation from some <001>//RD orientations such as {100}<001> and {110}<001> towards relatively stable <110>//RD (α-fiber) and <111>//ND orientations (γ-fiber) [5]. Thus, low finish rolling temperature can sharp hot rolled γ-fiber. During annealing, the sharp hot rolled γ-fiber is inclined to increase the number of γ-fiber recrystallized grains [5]. Hence, the low finish rolling temperature can decrease the fraction of the <001>//RD oriented grains. Because subsize Charpy specimens were machined along the RD, the annealed plates with low finish rolling temperature has lower fraction of grains with {001} planes which are parallel to notch plane than that with high finish rolling temperature.

Ghosh et al. [6] found that higher fraction of {001} planes which are parallel to the notch plane can promote the propagation of cleavage crack and eventually the increase of DBTT. The lower fraction of {001} planes parallel to the notch plane in the annealed plate LFT can decrease the DBTT. On the other hand, grain refinement can suppress the formation of deformation twins which can promote the initiation and propagation of crack during Charpy impact testing for FSSs [3]. The finer recrystallized microstructure obtained in the annealed plate LFT can decreases the DBTT. Therefore, the annealed plate LFT possesses lower the DBTT and low finish rolling temperature result in the decrease of the DBTT.

Figure 4 shows the microstructures of annealed plates with different annealing conditions analysed by EBSD. The average grain sizes of annealed plates annealed at 900, 950 and 1000°C are measured to be 92, 100 and 107μm, respectively (Figures 4(a), (c) and (e)). This indicates that the microstructure of annealed plates begins to coarsen with increasing annealing temperature. The fraction of <001>//RD oriented grains in the annealed plates annealed at 900, 950 and 1000°C are measured to be 20.5%, 18.7% and 17.1%, respectively (Figure 4(b), (d) and (f)), indicating that the fraction of grains with {001} planes which are parallel to notch plane in annealed plates decreases with the increasing annealing temperature. During annealing, the γ-fiber recrystallization grains preferentially nucleate and then these grains will grow by consuming deformed grains or other oriented recrystallization grains, such as <001>//RD oriented grains [5,7]. Increasing annealing temperature is favorable for increasing the driving force for recrystallization grain growth. Hence, increasing annealing temperature can reduce the fractions of <001>//RD oriented grains and {001} planes which are parallel to notch plane and coarsen recrystallized microstructure. As above-mentioned discussion, decreasing the fraction of grains with {001} planes which are parallel to notch plane can decrease the DBTT, while coarsening recrystallized microstructure can increase the DBTT. Therefore, the annealing temperature has no obvious influence on the DBTT.

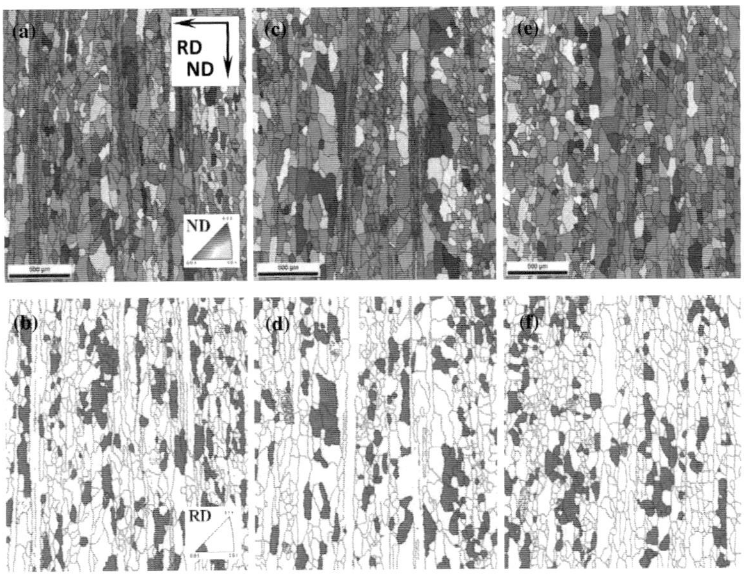

Figure 4 The orientation maps for all orientations ((a),(c),(e)) and corresponding <001>//RD oriented grains ((b),(d),(f)) measured from the annealed plates after different annealing processes (a),(b) 900°C, (c),(d) 950°C, (e),(f) 1000°C

4. Conclusions

The rolling and annealing processes have an obvious influence on microstructure and toughness for the developed AISI444 ferritic stainless steel having superior toughness in this study. Low finish rolling temperature could significantly refine the microstructure in hot rolled and annealed plate and decrease the volume fraction of the <001>//RD oriented grains and corresponding the fraction of grains with {001} planes which are parallel to notch plane in recrystallized microstructure. Therefore, the ductile-brittle transition temperature was lowered from -44.2°C to -48.8°C, and the toughness was enhanced after employing the low finish rolling temperature for developed AISI444 ferritic stainless steel. After annealing with higher temperature, the microstructure in hot rolled and annealed plate coarsened, but the volume fraction of the <001>//RD oriented grains and corresponding the fraction of grains with {001} planes which are parallel to notch plane in recrystallized microstructure decreased. Thus, the ductile-brittle transition temperature has no significant change when high annealing temperature is applied.

5. References

[1] Gao F, Liu Z Y, Liu H T and Wang G D 2013 Toughness under different rolling processes in ultra purified Fe–17 wt%Cr alloy steels J. Alloy. Compd. 567 141–47
[2] Gao F, Yu F X, Zhang X J, Li C G, Cao G M and Liu Z Y 2016 China Patent 2016 1 1062773.8

[3] Gao F, Yu F X, Misra R D K, Liu H T and Liu Z Y 2015 Correlation between ductile-to-brittle transition behaviour and twinning in ferritic stainless steel *Mater. Sci. Technol.* **31** 1462–72

[4] Han J, Li H J and Xu H G 2014 Microalloying effects on microstructure and mechanical properties of 18Cr–2Mo ferritic stainless steel heavy plates *Mater. Design* **58** 518–26

[5] Gao F, Liu Z Y, Liu H T and Wang G D 2013 Texture evolution and formability under different hot rolling conditions in ultra purified 17%Cr ferritic stainless steels *Mater. Charact.* **75** 93–100

[6] Ghosh A, Modak P, Dutta R and Chakrabarti D 2016 Effect of MnS inclusion and crystallographic texture on anisotropy in Charpy impact toughness of low carbon ferritic steel Mater. *Sci. Eng.* A 654 298–308

[7] Gao F, Yu F X, Misra R D K, Zhang X J, Zhang S M and Liu Z Y 2015 Microstructure, texture, and deep drawability under two different cold-rolling processes in ferritic stainless steel J. *Mater. Eng. Perform.* **24** 3862–80

Acknowledgments

The authors are grateful to the National Natural Science Foundation of China (No. 51704066), China Postdoctoral Science Foundation (2015M581346) and Foundation Research Funds for the Central Universities (N150203002).

AUTHOR INDEX